Lecture Notes in Artificial Intell

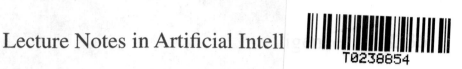

T0238854

Subseries of Lecture Notes in Computer Science

LNAI Series Editors

Randy Goebel
University of Alberta, Edmonton, Canada
Yuzuru Tanaka
Hokkaido University, Sapporo, Japan
Wolfgang Wahlster
DFKI and Saarland University, Saarbrücken, Germany

LNAI Founding Series Editor

Joerg Siekmann
DFKI and Saarland University, Saarbrücken, Germany

Igor Paprotny   Sarah Bergbreiter (Eds.)

Small-Scale Robotics

From Nano-to-Millimeter-Sized Robotic Systems
and Applications

First International Workshop at ICRA 2013
Karlsruhe, Germany, May 6, 2013
Revised and Extended Papers

 Springer

Volume Editors

Igor Paprotny
University of Illinois
Department of Electrical and Computer Engineering
Chicago, IL, USA
E-mail: paprotny@uic.edu

Sarah Bergbreiter
University of Maryland
Department of Mechanical Engineering
College Park, MD, USA
E-mail: sarahb@umd.edu

ISSN 0302-9743   e-ISSN 1611-3349
ISBN 978-3-642-55133-8   e-ISBN 978-3-642-55134-5
DOI 10.1007/978-3-642-55134-5
Springer Heidelberg New York Dordrecht London

Library of Congress Control Number: 2014936258

LNCS Sublibrary: SL 7 – Artificial Intelligence

Typesetting: Camera-ready by author, data conversion by Scientific Publishing Services, Chennai, India

Printed on acid-free paper

Springer is part of Springer Science+Business Media (www.springer.com)

Preface

In the 1968 movie The Fantastic Voyage, a team of scientists is reduced in size to micro-scale dimensions and embarks on an amazing journey through the human body, along the way interacting with human microbiology in an attempt to remove an otherwise inoperable tumor. Today, a continuously growing group of robotic researchers are attempting to build tiny robotic systems that perhaps one day can make the vision of such direct interaction with human microbiology a reality. This is in addition to numerous other applications ranging from micro- and nano-manufacturing to building tools for new scientific discoveries to search and rescue. This book contains selected contributions from some of the most renowned researchers in the field of small-scale robotics, based largely on invited presentations from the workshop "The Different Sizes of Small-Scale Robotics: from Nano-, to Millimeter-Sized Robotic Systems and Applications," which was held in conjunction with the IEEE/RAS International Conference on Robotics and Automation (ICRA 2013), in May 2013 in Karlsruhe, Germany. With many potential applications in areas such as medicine, manufacturing, or search and rescue, small-scale robotics represent a new emerging frontier in robotics research. The aim of this book is to provide an insight into ongoing research and future directions in this novel, continuously evolving field, which lies at the intersection of engineering, computer science, material science, and biology.

Igor Paprotny

About the Editors

Igor Paprotny is an Assistant Professor in the Department of Electrical and Computer Engineering at University of Illinois, Chicago, where he has been on the faculty since 2013. He has also been an Affiliated Scientist at the Lawrence Berkeley National Laboratory, Berkeley, CA since 2011, and the lead of the UIC/LBNL/UCB/EPA Air-Microfluidics Group. From 2008 to 2013 he was a post-doctorate and later a research scientist at the University of California, Berkeley. He received a Ph.D. in Computer Science in 2008 (Dartmouth College), BS and MS degrees in Industrial Engineering in 1999 and 2001 respectively (Arizona State University). He also holds an Engineering Diploma in Mechatronics (NKI College of Engineering, Oslo, 1995). His research interests include applications of micro electro mechanical systems (MEMS) to microrobotic, air-microfluidics, and low-cost non-intrusive energy systems monitoring.

Sarah Bergbreiter joined the University of Maryland, College Park in 2008 as an Assistant Professor of Mechanical Engineering, with a joint appointment in the Institute for Systems Research. She received her B.S.E. degree in Electrical Engineering from Princeton University in 1999, and the M.S. and Ph.D. degrees from the University of California, Berkeley in 2004 and 2007 with a focus on microrobotics. She received the DARPA Young Faculty Award in 2008, the NSF CAREER Award in 2011, and the Presidential Early Career Award for Scientists and Engineers (PECASE) in 2013 for her research on engineering robotic systems down to sub-millimeter size scales. She also received the Best Conference Paper Award at IEEE ICRA 2010 on her work incorporating new materials into microrobotics and the NTF Award at IEEE IROS 2011 for early demonstrations of jumping microrobots.

Table of Contents

Small-Scale Robotics : An Introduction

Igor Paprotny[1] and Sarah Bergbreiter[2]

[1] University of Illinois, Chicago IL 60607, USA
[2] University of Maryland, College Park, MD 20742, USA

Abstract. The term small-scale robotics describes a wide variety of miniature robotic systems, ranging from millimeter sized devices down to autonomous mobile systems with dimensions measured in nanometers. Unified by the common goal of enabling applications that require tiny mobile robots, research in small-scaled robotics has produced a variety of novel miniature robotic systems in the last decade. As the size of the robots scale down, the physics that governs the mode of operation, power delivery, and control change dramatically, restricting how these devices operate, and requiring novel engineering solutions to enable their functionality. This chapter provides an overview and introduction to small-scale robotics, drawing parallels to systems presented later in the book. Comparison to biological systems is also presented, using biology to speculate regarding future capabilities of robotic systems at the various size scales.

1 Introduction

The term small-scale robotics is used to describe smaller-than-conventional robotic systems, ranging from several millimeters to nanometers in size. Research in small-scale robotic systems is unified by the common goal of developing autonomous robotic machines for applications that require the individual robotic units to be of small (millimeters or smaller) size. Applications for such robots are numerous, including areas such as medicine, manufacturing, or search and rescue. However, robots at these scales must overcome many challenges related to their fabrication, control, and power delivery. This chapter provides a brief summary of the ongoing research in small-scale robotics, outlining some of the challenges related to development and implementation of robotic systems at these different scales. The following chapters of this book show several example implementations of selected small-scale robotic systems. They include selected papers based on presentations from the workshop "The Different Sizes of Small-Scale Robotics: from Nano-, to Millimeter-Sized Robotic Systems and Applications," which was held in conjunction with the International Conference on Robotics and Automation (ICRA 2013), in May 2013 in Karlsruhe, Germany.

Richard Feynman was perhaps the first to mention the idea of small-scale (non-mobile) robots in his 1959 lecture, "There's Plenty of Room at the Bottom," in which he described the possibility of tiny hands building even tinier hands [1]. Feynman followed this up with a lecture in 1983 titled "Infinitesimal Machinery" in which he mentioned the possibility of *mobile* small-scale robots for the first

I. Paprotny and S. Bergbreiter (Eds.): Small-Scale Robotics 2013, LNAI 8336, pp. 1–15, 2014.
© Springer-Verlag Berlin Heidelberg 2014

time [2]. He even proposed ideas for application of these robots including medical procedures and a game in which a millirobot could be used to fight a paramecium.

This decade coincided with the development of microelectromechanical systems (MEMS) which offered the ability to make the small motors, sensors, and mechanisms required for small robots for the first time. MEMS used the same fabrication processes that were developed for integrated circuits that could also be used for robot control. In 1987, Anita Flynn was the first to describe the possibilities enabled for robotics by MEMS fabrication methods in her paper on "Gnat Robots (And How They Will Change Robotics)" [3]. Flynn envisioned large numbers of small, inexpensive, and even disposable robots replacing a single large complex robot for a number of different tasks. Smart Dust systems conceived by Pister, Katz and Kahn can also be considered as a forerunner of small-scale robotics [4]. Smart dust nodes effectively include everything a mobile robot needs except for the actuation and propulsion. Operational milli-scale mobile robots composed of MEMS components were demonstrated in 2003 [5], and the first MEMS micro-scale mobile robots were developed in 2006 [6, 7]. Engineered passively floating autonomous nanoscale systems (i.e., engineered nanoparticles for drug delivery) started to emerge around 2000 (e.g., [8]), while molecular (electrochemical) walkers were demonstrated as early as 1991 [9]. Externally controlled propulsion and steering at the nanoscale has yet to be demonstrated.

Due to differences in fabrication techniques and operation principles, it is most convenient to classify small-scale robotic systems into size domains based on which units most consistently describe their size. Robotic Devices measured in millimeters are categorised as millirobots. devices smaller than a millimeter but larger than one micrometer are classified as microrobots, and devices smaller than one micrometer are classified as nanorobots. The size is determined by their largest dimension, i.e., a microrobot must fit within a cube 1 mm × 1 mm × 1 mm in size.

2 Millimeter-scale Systems

2.1 Overview

After Flynn's description of gnat robots in 1987 [3], it took over 10 years for the first robots at millimeter-size scales to become a reality. A significant reason for this delay was the challenge of making and assembling the components necessary for a millimeter-sized robot. Robots designed at centimeter-scales can generally rely on assembly of off-the-shelf components. For example inexpensive actuators (DC motors and shape memory alloy), sensors (light, inertial, etc.), and micro controllers can be combined to make both wheeled robots and legged robots [10, 11]. When approaching and moving below 1 centimeter characteristic lengths however, assembly of off-the-shelf components is no longer feasible. Millimeter-scale motors and mechanisms are generally not available off-the-shelf and must be fabricated. A millimeter-sized power source is required to efficiently power the motors and control circuitry. Controllers and sensors must be small and incredibly low power.

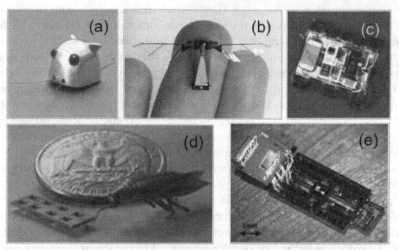

Fig. 1. Some previous mobile robots at sizes just over 1 cm and under. (a) Epson's Monsieur, from [12], (b) the Harvard Micro Robotic Insect, from [13] ©2008 IEEE, (c) a jumping millirobot using energetics, from [15] ©2012 IEEE, (d) Ebefors walking milli robot, from[16, 17], and (e) a 10 mg silicon walking millirobot, from [18].

The primary challenges in the development of mobile millirobots with a characteristic length < 1 cm include:

1. Mechanisms – How will robust mechanisms like legs and wings be fabricated at these scales?
2. Motors – What is the best way to design and fabricate efficient motors with high power densities to support locomotion?
3. Control – How can sensors and controllers be integrated on robots at this size scale and what are the best controllers to use for stable locomotion?
4. Power – What energy storage technologies can be integrated at these size scales and what power electronics are necessary to integrate with motors and controllers?
5. Mobility – How will these robots move through their environment? What methods of locomotion are effective and efficient at these size scales?

At larger sizes, millirobots can easily resemble their larger scale cousins. In fact, one of the first robots at this size scale was built by Seiko and used some of the motors and gears developed for watches [12]. While just over the maximum 1 cm characteristic length in the millirobot definition above, Monsieur (Fig. 1a) is one of the most autonomous robots at this size scale as it included sensing, power (in the form of a capacitor), and a reactive control system. Another slightly larger robot is the Harvard RoboBee [13, 14]. With a wingspan of approximately 3 cm, RoboBee (Fig. 1b) already requires novel mechanisms, actuators, and assembly to achieve the low mass and high output power necessary for controlled flight.

As envisioned earlier by Flynn, the most common approach to creating sub-centimeter robots has been the use of microfabrication and MEMS. Simoyama proposed the use of microfabrication to create insect-like exoskeletons [19] and

Yasuda implemented a robot using these ideas that scooted along a vibrating surface that provided the power required for locomotion [20]. While still larger than 1 cm, Ebefors demonstrated a 15 x 5 mm^2 millirobot that walked forward at speeds up to 6 mm/s on a smooth silicon wafer, carried 30 times its own weight, and survived relatively rough handling [17]. Ebefors' robot used polymer-based thermal actuators which led to significantly increased robustness, but a power consumption of over 1 W so it only operated while tethered to a power supply. Hollar integrated electrostatic motors, microfabricated and articulated legs, solar cells for power, and a CMOS open-loop controller to demonstrate autonomous millirobot locomotion with a characteristic length of 8.5 mm [18, 21]. While the robot did shuffle to one side, it was not able to walk forward as originally intended.

The primary focus for all of these microfabricated robots was the development of the mechanisms and actuators. The integration of sensing and control for a robot that 'senses, thinks, and acts' was largely neglected in these first millirobots. A prototype millirobot developed as part of the iSwarm project was one of the first to integrate communication, sensing, actuation, and control in a 3.9 x 3.9 x 3.3 mm^3 package [22]. While each of the components worked separately, the fully assembled robot did not demonstrate successful locomotion. Churaman used low-temperature solder assembly to integrate a light sensor, MOSFET, capacitors for power, and a microfabricated energetic material for jumping locomotion on a 7 x 4 x 4 mm^3 robot [15, 23]. This 300 mg robot jumped approximately 8 cm straight up in response to turning on a light. However, thus far, the robot has only included a one-time actuator and has only jumped once. There is clearly a significant amount of progress required to achieve the kind of mobile robots envisioned by Feynman and Flynn.

2.2   Applications

Millirobots are still at size scales at which small sensors can be included and fast (relative) locomotion can ultimately be achieved. After all, ants and other small insects already include numerous sensors and can run at speeds over 40 body lengths per second [24]. As such, these robots can be useful in a number of applications that large robots currently dominate in addition to many new ones. For example, survivors after an earthquake could be found faster if robotic bugs could target the efforts of first responders [25]. Such robots could also provide low-cost sensor deployment over civil infrastructure [26] and engage in stealthy surveillance [27].

New applications enabled by millirobots include scientific instrumentation, highly parallel manufacturing, and robots moving inside the human body for medical interventions. A number of millirobots at the top end of the size constraints have targeted applications in science [28, 29, 22]. Robots that include their own scanning probe microscopes (SPMs) or atomic force microscopes (AFMs) can be used to study new materials or biological samples along with other micro and nano scale features. Large numbers of small mobile robots could also be used for manufacturing – one of the goals envisioned by Feynman for small-scale robots [1] and one of the original inspirations for MEMS [30]. Ants and termites build impressive mounds and millirobots could be used to manufacture large structures in a similar fashion [31]. Instead of using given materials for manufacturing, one millirobot targets manufacturing by using itself as the building material [32]. These 'catoms' provide the basic building blocks for 'claytronics' – a system in which thousands to millions of robots can reassemble themselves into useful shapes [33]. In medicine, robots at sub-cm size scales also offer new opportunities for evaluation, surgery, and monitoring. Imagine a pill camera that could grab a biopsy as it passes by a tumor or one of other the numerous (and currently larger) robots designed to navigate the gastrointestinal tract [34]. These robots could also be used to explore, manipulate, and repair the interstitial spaces in the human body [35].

2.3 Biological Comparison

A clear inspiration for mobile millirobots is the huge variety of animal species that fits within these same size constraints. Previous research in millirobotics has never come close to achieving or explaining the remarkable mobility seen in insects. Ants (which make up 15-20% of the entire terrestrial animal biomass [36]) can move at speeds over 40 body lengths/second and carry loads greater than 4 times their own body mass on a wide variety of surfaces [37, 24]. Jumping insects like froghoppers can jump as high as 700 mm with accelerations over 400 g [38]. Flying insects like *Drosophila* (the common fruit fly) include incredible sensing and maneuverability in flight [39]. Underwater, copepods (among others) both drift or actively locomote through their environments [40]. Copepods are found in the ocean and nearly every freshwater habitat.

All of these organisms display impressive autonomy that is still missing in the millirobot examples mentioned above. They clearly display power autonomy and are able to refuel for lifespans that last from days to years. They exhibit impressive efficiency during locomotion as well. By measuring the oxygen consumption of ants like *Camponotus*, it has been shown that the entire ant burns approximately 140 μW while walking on a flat surface [41]. Such efficient and effective locomotion used by insects at slightly larger scales (cockroaches) has been used as inspiration for high speed locomotion in larger running robots like RHex and iSprawl [42, 43]. Finally, many of these animals display impressive sensing throughout their bodies – *drosophila* especially displays impressive visual sensing during flight that is copied on much larger robots [44]. This sensing is used for impressive control and maneuverability as well – for example corrective turns in 30 msec for *Drosophila* [39].

3 Microscale Systems

3.1 Overview

Due to further miniaturization of components and corresponding challenges associated with fabrication of robotic systems at the microscale, microrobotic systems lack the complexity of their millimeter-sized counterparts. Surface forces dominate, and thus volume-based energy storage (e.g., batteries) is less effective. As a result, most microrobotic systems employ external power delivery. Because microrobotic systems are build by extending untethered microactuator mechanisms, they can be classified according to the underlying propulsion method. The most common mechanisms include electrostatic actuation [45, 6, 46], magnetic actuation [7, 47–50], as well as biologic propulsion using motile bacteria (e.g., [51, 52]. An excellent review of biologic propulsion is presented here [53]). Catalytic microengines [54] and optically actuated microsystems [55, 56] are among other promising areas of future microrobotic propulsion. Four selected microrobotic systems are shown in Figure 2.

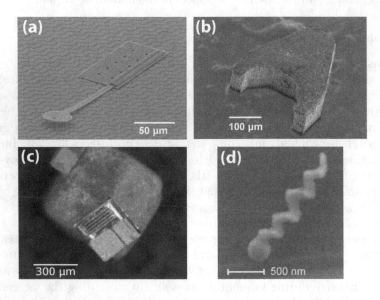

Fig. 2. Different types of mobile microrobots, including electrostatic (a) and magnetically driven (b-d) microrobots. Images obtained with permission from [6] (a), [47] (b), [49] (c), and [48] (d).

Due to integration challenges as well as the lack of an on-board power source, microrobotic systems are at present largely controlled via teleoperation, i.e. the devices are controlled through a global control signal. On-board control and sensing has up to this date not been implemented, with the exception of biologically propelled systems using reactive control, e.g., magnetotaxis [51] or chemotaxis [52] of *Escheria coli*. Simple compliant feedback control, a form of tactile sensing was also implemented by [57]. On-board sensing and control of non-biologically propelled microrobots is an active area of future research.

The requirement to control many simple devices using a single (or at best a limited number of) global control signal necessitates novel approaches to multi-microrobot control. These include global control selective response (GCSR) [58] and Ensemble Control [59]. Both of these approaches *engineer* the trajectories of several simultaneously moving robots by taking advantage of their motion during the application of the global control signal. GCSR has been successfully used to control multiple microrobots to achieve microassembly [57].

3.2 Applications

Applications of microbotic systems are numerous. In medicine potential tasks include non-intrusive surgery [7], target drug delivery, and *in-vivo* tissue engineering [60]. The ability to construct, i.e. assemble, microstructures (factory on a chip) could help to facilitate bottom-up assembly of complex microsystems [61, 58, 62]. Microrobots could be envisioned to maintain the structure, change the shape to fit the task at hand, and heal it when it becomes damaged. Other exciting applications include surveillance and information security, where the mere size of microrobots makes them nearly undetectable. Finally, microrobotics as a field is inspiring a new generation of engineers, for example through events such as the Mobile Microrobotic Challenge [63].

3.3 Biological Comparison

Whereas robotic systems by and large have to rely on external delivery of power and control as their dimensions are reduced to the microscale, biological systems remain fully autonomous i.e., contain on-board control, power storage, sensing and actuation, down to the size of a few hundred μm. For example, a dust mite (*Dermatophagoides farinae*) [64] is only 400 μm long, while an amoeba (e.g., *Amoeba proteus*) [65] is between 200 μm and 700 μm in size. The smallest insect on Earth is the male fairy fly (*Dicopomorpha zebra*) [66]. Although the thorax of a female fairy fly measures a few hundred μm, the male of that species is only approximately 130 μm in length. It is blind and no longer mobile, its goal is to attach itself to a female with the sole purpose of fertilizing her eggs. The male fairy fly marks the point where biological systems become less autonomous. This reduction in autonomy is presumably caused by the absence of a central nervous systems, which coincides with the transition to single-cellular organisms. A human sperm-cell is 55 μm long (although its head measures only 3-5 μm), an E-coli bacteria (*Escherichia coli*) measures only 2 μm in length. Both are capable of autonomous propulsion, however the control is limited to following simple chemical gradients (i.e. purely reactive control).

4   Nanoscale Systems

4.1   Overview

Transition to sub-micrometer dimensions further compounds the challenges associated with fabrication and control of autonomous systems. Autonomous mobile nanoscale systems can be divided into two categories: passive untethered systems or molecular walkers.

Untethered nanoscale devices moving through fluid would experience very high viscous drag due to extremely low Reynolds numbers [67], and controlled propulsion becomes difficult. As a consequence, passive untethered nanoscale systems rely on random motion of surrounding fluid for propulsion. Nanoparticles laden with chemotherapy drugs are examples of systems that are able to selectively enter cancerous cells, detect the point of entry, and release their payload once inside the cancerous cell. These relatively simple nano-particles are able to accomplish all this in the absence of on-board propulsion or control [8, 68–70]. They are already showing great promise in the treatment of certain aggressive forms of cancer [68].

Apart from drug delivery, random motion and stochastic interaction is also used in molecular self-assembly (MSA) where random mutual interaction of many units and selective affinity results in the generation (assembly) of shapes [71–73], or computing constructs [74]. In these systems, the application relies on the dispensation of a very large number of units, such that the random process responsible for actuation results in a sufficient number of them reaching the goal state.

The second class of autonomous nanoscale systems are molecular walkers, which are directed chemical actuators capable of controlled propulsion along a substrate. Molecular shuttles, i.e. molecular machines capable of carrying molecular cargo along a larger molecule or a chain of molecules, were initially demonstrated by Anelli et al [9]. Limited control of molecular transport by these shuttles was demonstrated by [75, 76]. In particular, the motor protein kinesin [77] is often used as a construct for the molecular shuttles. At these scales, the boundary between chemical and mechanical interaction, as well as biological and artificial constructs becomes somewhat fluid. Shirai et al. [78] demonstrated controlled rolling motion of a C_{60} fullerene (buckyballs) molecular car actuated by an STM tip. A more advanced version of molecular shuttles following tracks on an engineered DNA origami substrate was shown by Lund et al. [79]. Although capable of following complex engineered tracks, molecular shuttles are at this point a practical actuation mechanism, and an efficient external control scheme must be developed before these systems can be classified as truly autonomous. Examples of the two types of autonomous nanoscale systems described above are shown Figure 3.

4.2 Biological Comparison

Biological mobile nanoscale systems use similar modes of operation as their engineered counterparts, which should be no surprise as the engineered systems often use biologically derived building blocks as their constructs. Hence, biological systems can also be divided into passive untethered systems and naturally occurring molecular walkers.

Viruses are perhaps the most commonly occurring untethered mobile biological nanoscale systems [80]. With sizes ranging from hundreds to tens of nanometers, these systems are strands of protein that seek out and attach to the receptors of the target cells, and use cellular machinery for replication. They

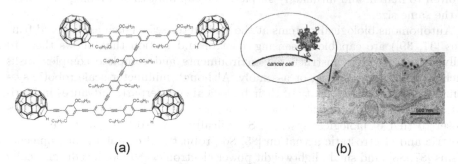

Fig. 3. Nanoscale autonomous systems. (a) An illustration of a molecular walker called *the molecular car* actuated by an scanning tunnelling microscope (STM) tip. (b) Engineered targeted nanoparticles entering a cancerous cell. Inset shows a cartoon of the particle with the targeting component interacting with the cancer cell interface prior to entry. Images reprinted from [78] (a) and [68] (b), with permission.

contain no on-board propulsion mechanisms, and are completely passive and opportunistic; they rely on the motion of the surrounding fluid rather than controlled propulsion to seek out the matching receptors of the host cell. Because their motion is random, they rely on numerous copies to ensure that that a sufficient concentration reaches the host cell. Viruses are very common, as they are wide-spread and are a cause for seasonal concern around the world (i.e., [81]).

Similarly, naturally occurring molecular walkers inside the cells of living organisms are instrumental for the transport of nutrients and proteins between the cell periphery and its core, as well as DNA replication and protein synthesis. Kinesin [77] is a naturally occurring biological walker which is responsible for the transport of larger molecules to and from the cellular interior, as well as possibly cellular mytosis [82]. To move molecular cargo, Kinesin strands walk along microtubule filament of the cellular cytoskeleton. Kinesin is unidirectional, and certain types of Kinesin walk toward the cell periphery while other types walk towards the cell center. Other types of molecular walkers can be found supporting DNA transcription and RNA synthesis. For example, DNA polymerase, which traverses the DNA strand and creates a complementary RNA strand [83]. Repair enzymes, such as DNA ligase [84], are involved in traversing the transcribed DNA-material and correcting the errors in chromosome transcription. Clearly, these naturally-occurring molecular walkers are present in every cell in every living organism on our planet, and are responsible for continuous operation of the cells. Their importance in maintaining the life on our planet cannot be overstated.

5   The Biology vs. Robotics Gap

The capabilities of small-scale robotic systems are regularly compared to that of biological systems of comparable size. The goal of small-scale robotics research is often to match, and ultimately surpass, the capabilities of a biological system of the same size.

Autonomous biological systems at the millimeter scale (such as ants and fruit flies [37, 39]) are capable of sensing, control and motion that allows them to fully traverse highly unstructured environments and complete complex tasks such as foraging, mapping, or assembly. Although millimeter scale robotic systems still lack the complexity of their biological counterparts, advances in fabrication and integration technologies are progressively bringing their capabilities closer to that of biological systems. Specifically, new work in low power piezoelectric and electrostatic actuation [85, 86], robust and compliant micromechanisms [87, 88], and small, lightweight power electronics [89, 90] has dramatically improved locomotion prospects for millirobots (although this mobility has yet to be demonstrated). The remarkable sensing, control, and integration seen in millimeter-scale organisms motivates future research in this area.

Remarkably, biological systems continue to exhibit highly autonomous behavior down to the size for a few hundred micrometers. For example, the 400 μm dust mite [64] is capable of autonomously navigating in search for food and traversing highly unstructured environments. Similar capabilities can be found in *Amobea proteous* or *Dicopomorpha zebra*. These fully autonomous capabilities of microscale biological systems stand in stark contrast with present capabilities of microrobotic systems. At present, microrobotic systems have at best only limited autonomy, lacking both on-board power generation and control (e.g., [6, 47, 48, 50]). In all of these cases, power and a control signal is provided by an off-board system, and the robots are essentially navigated through teleoperation, with the exception of biologically powered systems such as [51, 52] that contain their own power generation and simple control. It is clear that compared to biological systems, microrobots at the several hundred micrometer scale still have much room to grow compared with their biological counterparts. This apparent gap should be viewed as motivation for future microrobotic research.

Further reduction in size below single micrometers transitions to the nanoscale size domain and results in biological systems that are either floating passively (such as viruses or phage) or are traversing a substrate (e.g. molecular walkers). At the nanoscale, the capabilities of engineered autonomous systems, or nanorobotics as defined in this chapter, are converging. This is perhaps not that surprising, as these systems have generally been engineered using biologically-derived building blocks.

6   Conclusions

In this chapter we have provided a summary of ongoing research in small-scale robotics and their applications, and several of the systems mentioned in this chapter are further described in detail in the subsequent chapters of this book. We also presented a comparison outlining the contrast between the current capabilities of small-scale robotic systems and biological systems of comparable scale. For the most part, biology is still far ahead of small-scale robotics in terms of autonomy and task capabilities. However, these gaps highlight important areas of research while demonstrating the level of autonomy that should be attainable by future robotic systems at all scales.

References

1. Feynman, R.P.: There's plenty of room at the bottom. Journal of Microelectrome-chanical Systems 1(1), 60–66 (1992)
2. Feynman, R.P.: Infinitesimal machinery. Journal of Microelectromechanical Systems 2(1), 4–14 (1993)
3. Flynn, A.M.: Gnat robots (and how they will change robotics). In: IEEE Micro Robots and Teleoperators Workshop: An Investigation of Micromechanical Structures, Actuators and Sensors, Hyannis, MA (1987)
4. Kahn, J.M., Katz, R.H., Katz, Y.H., Pister, K.S.J.: Emerging challenges: Mobile networking for "smart dust". Journal of Comunications and Networks 2, 188–196 (2000)
5. Hollar, S., Flynn, A., Bellew, C., Pister, K.S.J.: Solar powered 10 mg silicon robot. In: The Proceedings of the the Sixteenth Annual International Conference on Micro Electro Mechanical Systems, MEMS 2003, Kyoto, Japan, January 19-23, pp. 706–711 (2003)
6. Donald, B.R., Levey, C.G., McGray, C., Paprotny, I., Rus, D.: An untethered, electrostatic, globally-controllable MEMS micro-robot. Journal of Microelectrome-chanical Systems 15(1), 1–15 (2006)
7. Yesin, K.B., Vollmers, K., Nelson, B.J.: Actuation, sensing, and fabrication for in vivo magnetic microrobots. In: Ang Jr., M.H., Khatib, O. (eds.) Experimental Robotics IX. STAR, vol. 21, pp. 321–330. Springer, Heidelberg (2006)
8. Majeti, N.V., Kumar, R.: Nano and microparticles as controlled drug delivery devices. Journal of Pharmaceutical Sciences 3(2), 234–258 (2000)
9. Anelli, P.L., Spencer, N., Stoddart, J.F.: A molecular shuttle. Journal of Americal Chemical Society 113(13), 5131–5133 (1991)
10. Sabelhaus, A.P., Mirsky, D., Hill, M., Martins, N.C., Bergbreiter, S.: TinyTeRP: A tiny terrestrial robotic platform with modular sensing capabilities. In: IEEE International Conference on Robotics and Automation, karisruhe, Germany (May 2013)
11. Hoover, A.M., Steltz, E., Fearing, R.S.: RoACH: an autonomous 2.4g crawling hexapod robot. In: IEEE/RSJ International Conference on Intelligent Robots and Systems, Nice, France (September 2008)
12. Epson: Monsieur: The ultraminiature robot that propelled itself into the Guinness Book (March 1993),
http://www.epson.co.jp/e/company/milestones_23_monsieur.htm
13. Wood, R.J.: The first takeoff of a biologically inspired at-scale robotic insect. IEEE Transactions on Robotics 24(2), 341–347 (2008)
14. Ma, K.Y., Chirarattananon, P., Fuller, S.B., Wood, R.J.: Controlled flight of a biologically inspired, insect-scale robot. Science 340(6132), 603–607 (2013)
15. Churaman, W.A., Currano, L.J., Morris, C.J., Rajkowski, J.E., Bergbreiter, S.: The first launch of an autonomous thrust-driven microrobot using nanoporous energetic silicon. Journal of Microelectromechanical Systems 21(1), 198–205 (2012)
16. Ebefors, T., Asplund, T.: A selection of photos of the walking micro-robot (January 2014), http://www.s3.kth.se/mst/research/gallery/microrobot_photo.html

17. Ebefors, T., Mattsson, J.U., Kälvesten, E., Stemme, G.: A walking silicon micro-robot. In: Transducers, Sendai, Japan, pp. 1202–1205 (1999)
18. Hollar, S., Flynn, A.M., Bellew, C., Pister, K.S.J.: Solar powered 10 mg silicon robot. In: IEEE Micro Electro Mechanical Systems, pp. 706–711 (2003)
19. Shimoyama, I., Miura, H., Suzuki, K., Ezura, Y.: Insect-like microrobots with external skeletons. IEEE Control Systems Magazine 13(1), 37–41 (1993)
20. Yasuda, T., Shimoyama, I., Miura, H.: Microrobot actuated by a vibration energy field. Sensors and Actuators A Physical 43, 366–370 (1994)
21. Hollar, S., Flynn, A.M., Bergbreiter, S., Pister, K.S.J.: Robot leg motion in a planarized-SOI, Two-Layer Poly-Si process. Journal of Microelectromechanical Systems 14(4), 725–740 (2005)
22. Edqvist, E., Snis, N., Mohr, R.C., Scholz, O., Corradi, P., Gao, J., Dieguez, A., Wyrsch, N., Johansson, S.: Evaluation of building technology for mass producible millimetre-sized robots using flexible printed circuit boards. Journal of Micromechanics and Microengineering 19(7), 075011(11pp) (2009)
23. Churaman, W.A., Gerratt, A.P., Bergbreiter, S.: First leaps toward jumping microrobots. In: IEEE/RSJ International Conference on Intelligent Robots and Systems, San Francisco, CA (September 2011)
24. Zollikofer, C.: Stepping patterns in ants - influence of speed and curvature. Journal of Experimental Biology 192(1), 95–106 (1994)
25. Kumar, V., Rus, D., Singh, S.: Robot and sensor networks for first responders. Pervasive Computing, 24–33 (October 2004)
26. Zhu, D., Qi, Q., Wang, Y., Lee, K.M., Foong, S.: A prototype mobile sensor network for structural health monitoring. In: Nondestructive Characterization for Composite Materials, Aerospace Engineering, Civil Infrastructure, and Homeland Security, vol. 7294 (April 2009)
27. Gage, D.W.: How to communicate with zillions of robots. In: SPIE Mobile Robots VIII, Boston, vol. 2058, pp. 250–257 (1993)
28. Driesen, W., Varidel, T., Regnier, S., Breguet, J.M.: Micro manipulation by adhesion with two collaborating mobile micro robots. Journal of Micromechanics and Microengineering 15(10), S259–S267 (2005)
29. Martel, S.: Fundamental principles and issues of high-speed piezoactuated three-legged motion for miniature robots designed for nanometer-scale operations. The International Journal of Robotics Research 24(7), 575–588 (2005)
30. Trimmer, W.: Microrobots and micromechanical systems. Sensors and Actuators 19(3), 267–287 (1989)
31. Werfel, J.: Anthills built to order: Automating construction with artificial swarms. PhD dissertation, Massachusetts Institute of Technology (May 2006)
32. Karagozler, M.E., Goldstein, S.C., Reid, J.R.: Stress-driven MEMS assembly + electrostatic forces = 1mm diameter robot. In: 2009 IEEE/RSJ International Conference on Intelligent Robots and Systems, St. Louis, MO, USA, pp. 2763–2769 (October 2009)
33. Goldstein, S.C., Mowry, T.C.: Claytronics: A scalable basis for future robots. In: RoboSphere 2004, Moffett Field, CA (November 2004)
34. Ciuti, G., Menciassi, A., Dario, P.: Capsule endoscopy: From current achievements to open challenges. IEEE Reviews in Biomedical Engineering 4, 59–72 (2011)
35. Platt, S., Hawks, J., Rentschler, M.: Vision and task assistance using modular wireless in vivo surgical robots. IEEE Transactions on Biomedical Engineering 56(6), 1700–1710 (2009)
36. Schultz, T.R.: In search of ant ancestors. Proceedings of the National Academy of Sciences 97(26), 14028–14029 (2000)

37. Zollikofer, C.: Stepping patterns in ants - influence of load. Journal of Experimental Biology 192(1), 119–127 (1994)
38. Burrows, M.: Froghopper insects leap to new heights. Nature 424, 509 (2003)
39. Frye, M.A., Dickinson, M.H.: Fly flight. Neuron 32(3), 385–388 (2001)
40. Andersen Borg, C.M., Bruno, E., Kiørboe, T.: The kinematics of swimming and relocation jumps in copepod nauplii. Plos One 7(10), e47486 (2012)
41. Lipp, A., Wolf, H., Lehmann, F.: Walking on inclines: energetics of locomotion in the ant camponotus. The Journal of Experimental Biology 208, 707–719 (2005)
42. Saranli, U., Buehler, M., Koditschek, D.E.: RHex: a simple and highly mobile hexapod robot. International Journal of Robotics Research 20, 616–631 (2001)
43. Kim, S., Clark, J.E., Cutkosky, M.R.: iSprawl: design and tuning for high-speed autonomous open-loop running. The International Journal of Robotics Research 25(9), 903–912 (2006)
44. Zufferey, J.C., Klaptocz, A., Beyeler, A., Nicoud, J.D., Floreano, D.: A 10-gram vision-based flying robot. Advanced Robotics 21, 1671–1684 (2007)
45. Donald, B.R., Levey, C.G., McGray, C., Rus, D., Sinclair, M.: Power delivery and locomotion of untethered micro-actuators. Journal of Microelectromechanical Systems 10(6), 947–959 (2003)
46. Valencia, M., Atallah, T., Castro, D., Conchouso, D., Dosari, M., Hammad, R., Rawashdeh, E., Zaher, A., Kosel, J., Foulds, I.G.: Development of untethered SU-8 polymer scratch drive microrobots. In: Proceedings of the 24th International Conference on Micro Electro Mechanical Systems (IEEE MEMS 2011), pp. 1221–1224 (January 2011)
47. Floyd, S., Pawashe, C., Sitti, M.: An untethered magnetically actuated micro-robot capable of motion on arbitrary surfaces. In: The Proceedings of IEEE International Conference on Robotics and Automation, ICRA (2008)
48. Frutiger, D.R., Kratochvil, B.E., Vollmers, K., Nelson, B.J.: Magmites wireless resonant magnetic microrobots. In: The Proceedings of IEEE International Conference on Robotics and Automation, ICRA (May 2008)
49. Ghosh, A., Fischer, P.: Controlled propulsion of artificial magnetic nanostructured propellers. Nano Letters 9(6), 2243–2245 (2009)
50. Jing, W., Pagano, N., Cappelleri, D.: A novel micro-scale magnetic tumbling microrobot. Journal of Micro-Bio Robotics 8(1), 1–12 (2013)
51. Martel, S.: Controlled bacterial micro-actuation. In: Proc. of the Int. Conf. on Microtechnologies in Medicine, pp. 89–92 (May 2006)
52. Kim, D., Liu, A., Diller, E., Sitti, M.: Chemotactic steering of bacteria propelled microbeads. Biomedical Microdevices 14(6), 1009–1017 (2012)
53. Martel, S.: Bacterial microsystems and microrobots. Biomedical Microdevices 14, 1033–1045 (2012)
54. Sanchez, S., Solovev, A.A., Harazim, S.M., Deneke, C., Mei, Y.F., Shmidt, O.G.: The smallest man-made jet engine. The Chemical Record 11(6), 367–370 (2011)
55. Chiou, P.Y.: Massively Parallel Optical Manipulation of Single Cells, Micro- and Nano-particles on Optoelectronic Devices. PhD thesis, University of California, Berkeley (2005)
56. Erb, R.M., Jenness, N.J., Clark, R.L., Yellen, B.B.: Towards holonomic control of janus particles in optomagnetic traps. Advanced Materials 21, 1–5 (2009)
57. Donald, B.R., Levey, C.G., Paprotny, I.: Planar microassembly by parallel actuation of MEMS microrobots. Journal of Microelectromechanical Systems 17(4), 789–808 (2008)

58. Donald, B.R.: Building very small mobile micro robots. Inaugural Lecture, Nanotechnology Public Lecture Series. (MIT (Research Laboratory for Electronics, Electrical Engineering and Computer Science, and Microsystems Technology, Laboratories), Cambridge (2007), http://mitworld.mit.edu/video/463/

59. Becker, A.T.: Ensemble Control of Robotic Systems. PhD thesis, University of Illinois at Urbana-Champaign (2012)

60. Khademhosseini, A., Langer, R., Borstein, J., Vacanti, J.P.: Microscale technologies for tissue engineering and biology. Proceedings of the National Academy of Science 103(8), 2480–2487 (2006)

61. Popa, D., Stephanou, H.E.: Micro and meso scale robotic assembly. SME Journal of Manufacturing Processes 6(1), 52–71 (2004)

62. Donald, B.R., Levey, C., Paprotny, I., Rus, D.: Planning and control for microassembly using stress-engineered. International Journal of Robotics Research 32(2), 218–246 (2013)

63. Popa, D., Cappelleri, D., Paprotny, I.: Mobile microrobotic challenge 2014 (2013), http://www.uta.edu/ee/ngs/mmc/rules.pdf

64. Lyon, W.F.: Ohio state university extension fact sheet: house dust mites (1991), http://ohioline.osu.edu/hyg-fact/2000/2157.html

65. Nishiharra, E., Shimmen, T., Sonobe, S.: Functional characterization of contractile vacuole isolated from amoeba proteus. Cell Structure and Function 29(4), 85–90 (2004)

66. Huber, J.T.: The genus dicopomorha (hymenoptera, mymaridae) in Africa and a key to alaptus-group genera. ZooKeys 20, 233–244 (2009)

67. Purcell, E.M.: Life at low reynolds number. American Journal of Physics 45, 3–11 (1977)

68. Davis, M.E.: The first targeted delivery of siRNA in humans via a self-assembling, cyclodextrin polymer-based nanoparticle: From concept to clinic. Molecular Pharmaceutics 6(3), 659–668 (2009)

69. Choi, C.H.J., Zuckerman, J.E., Webster, P., Davis, M.E.: Targeting kidney mesangium by nanoparticles of defined size. Proceedings of the National Academy of Science 108(16), 6656–6661 (2011)

70. Zuckerman, J.E., Choi, C.H.J., Han, H., Davis, M.E.: Polycation-siRNA nanoparticles can disassemble at the kidney glomerular basement membrane. Proceedings of the National Academy of Science 109(8), 3137–3142 (2012)

71. Seeman, N.C.: Nucleic acid nanostructures and topology. Angewandte Chemie International Edition 37(23), 3220–3238 (1998)

72. Whitesides, G.M., Grzybowski, B.: Self-assembly at all scales. Science 295, 2418–2421 (2002)

73. Rothemund, P.W.K.: Folding dna to create nanoscale shapes and patterns. Nature 446, 297–302

74. Adleman, L.M.: Molecular computation of solutions to combinatorial problems. Science 266(5187), 1021–1024 (1994)

75. Bissell, R., Cordova, E., Kaifer, A.E., Stoddart, J.F.: A chemically and electrochemically switchable molecular shuttle. Nature 369, 133–137 (1994)

76. Hess, H., Clemmens, J., Qin, D., Howard, J., Vogel, V.: Light-controlled molecular shuttles made from motor proteins carrying cargo on engineered surfaces. Nano Letters 1(5), 235–239 (2001)

77. Vale, R.D., Reese, T.S., Sheetz, M.P.: Identification of a novel force-generating protein, kinesin, involved in microtubule-based motility. Cell 42(1), 39–50 (1985)

78. Shirai, Y., Osgood, A.J., Zhao, Y., Kelly, K.F., Tour, J.M.: Directional control in thermally driven single-molecule nanocars. Nano Letters 5(11), 2330–2334 (2005)

79. Lund, K., Manzo, A.J., Dabby, N., Michelotti, N., Johnson-Buck, A., Nagreave, J., Taylor, S., Pei, R., Stojanovic, M.N., Walter, N.G., Winfree, E., Yan, H.: Molecular robots guided by prescriptive landscapes. Nature 465, 206–210 (2010)
80. McGrath, S., Sinderen, D.: Bacteriophage: Genetics and Molecular Biology. Caister Academic Press (2007)
81. Centers for Disease Control and Prevention: Seasonal inlueanza, flu (2013), http://www.cdc.gov/flu/ (downloaded on December 30, 2013)
82. Goshima, G., Vale, R.D.: Cell cycle-dependent dynamics and regulation of mototic kinesins in drosophila s2 cells. Molecular Biology of the Cell 16(8), 3896–3907 (2005)
83. Lehman, I.R., Bessman, M.J., Simms, E.S., Kornberg, A.: Enzymatic synthesis of deoxyribonucleic acid. i. Preparation of substrates and partial purification of an enzyme from escheria coli. Journal of Biological Chemistry 233(1), 163–170 (1958)
84. Lehman, I.R.: Dna ligase: structure, mechanism, and function. Science 186(4166), 790–797 (1974)
85. Pulskamp, J.S., Polcawich, R.G., Rudy, R.Q., Bedair, S.S., Proie, R.M., Ivanov, T., Smith, G.L.: Piezoelectric PZT MEMS technologies for small-scale robotics and RF applications. MRS Bulletin 37(11), 1062–1070 (2012)
86. Penskiy, I., Bergbreiter, S.: Optimized electrostatic inchworm motors using a flexible driving arm. Journal of Micromechanics and Microengineering 23(1), 015018 (2013)
87. Gerratt, A.P., Bergbreiter, S.: Incorporating compliant elastomers for jumping locomotion in microrobots. Smart Materials and Structures 22(1), 014010 (2013)
88. Whitney, J.P., Sreetharan, P.S., Ma, K.Y., Wood, R.J.: Pop-up book MEMS. Journal of Micromechanics and Microengineering 21(11), 115021 (2011)
89. Tang, Y., Chen, C., Khaligh, A., Penskiy, I., Bergbreiter, S.: An ultra-compact dual-stage converter for driving electrostatic actuators in mobile microrobots. IEEE Transactions on Power Electronics 29(6), 2991–3000 (2014)
90. Karpelson, M., Wei, G.Y., Wood, R.J.: Driving high voltage piezoelectric actuators in microrobotic applications. Sensors and Actuators A: Physical 176, 78–89 (2012)

Tubular Micro-nanorobots: Smart Design for Bio-related Applications

Samuel Sánchez[1,2], Wang Xi[1,2], Alexander A. Solovev[1], Lluís Soler[1,2],
Veronika Magdanz[2], and Oliver G. Schmidt[2,3]

[1] Max Planck Institute for Intelligent Systems, Heisenbergstr. 3, 70569 Stuttgart, Germany
[2] Institute for Integrative Nanosciences, IFW Dresden, Helmholtzstrasse 20,
D-01069 Dresden, Germany
[3] Material Systems for Nanoelectronics, Chemnitz University of Technology,
Reichenhainer Strasse 70, 09107 Chemnitz, Germany
sanchez@is.mpg.de

Abstract. We designed microrobots in the form of autonomous and remotely guided microtubes. One of the challenges at small scales is the effective conversion of energy into mechanical force to overcome the high viscosity of the fluid at low Reynolds numbers. This can be achieved by integration of catalytic nano-materials and processes to decompose chemical fuels. However, up to now, mostly hydrogen peroxide has been employed as a fuel which renders the potential applications in biomedicine and *in vivo* experiments. Therefore, other sources of energy to achieve motion at the micro- nanoscale are highly sought-after. Here, we present different types of tubular micro- and nanorobots, alternative approaches to toxic fuels and also, steps towards the use of tubular microrobots as micro- and nanotools.

Keywords: rolled-up, nanorobotics, self-propulsion, jet engines, autonomous.

1 Introduction

Motor proteins and biological micro- and nano-machines represent clear inspiration for engineers to design artificial motile robots of different sizes, shapes and aimed for multiple tasks. Ideally, these man-made robots should take the energy from the surroundings, sense their environment and be able to perform complex tasks in a similar way biomotors do.[1] Looking at nature, biological motors are powered by chemical fuels, efficiently generating propulsion power from chemical energy in an autonomous manner and by employing catalytic reactions.[2] Those living organisms not only move autonomously but also can react to external stimuli to orient or protect themselves. Bio-mimicry of chemo- and phototaxis have been recently demonstrated by catalytic micromotors.[3]

Locomotion in the regime of low Reynolds numbers is challenging task to be tackled when developing nano-and microrobotics. Since the pioneering works on synthetic catalytic nanomotors,[4,5,6] considerable efforts have been put towards the efficient conversion of chemical energy into mechanical one, addressing challenges of artificial micromachines to reach the comparable performance of highly efficient

I. Paprotny and S. Bergbreiter (Eds.): Small-Scale Robotics 2013, LNAI 8336, pp. 16–27, 2014.

biological nanomotors.[7,8] Various propulsion mechanisms can be employed to achieve motion of catalytic microbots and several methods of motion control have been reported.[9] Previously, toxic fuels have been used for microbots, such as hydrazine and high concentration of hydrogen peroxide[10] which unfortunately renders the use of artificial microbots in real biological samples nearly impossible. Recently, other alternatives have been proposed and several groups are working on the finding of more biocompatible fuels. If hydrogen peroxide fuel is used for powering catalytic microbots, the main reaction occurring at the surfaces of the microrobots is the breakdown of H_2O_2 into O_2 and H_2O, as has been recently reviewed by several groups.[11,12] Tubular microjet engines based on rolled-up nanotech[13] present some advantages compared to other man-made nanomotors such as larger surface area, on-demand tuning of materials' properties and device dimensions, mass production and reproducibility.[14,15,16] In addition, the microtubular jet engines rely on the microbubble propulsion mechanism which is not affected by the ionic strength of the solution.[17] Tubular microrobots can contain magnetic thin layers which enable the magnetic remote guidance with relatively high accuracy, leading to a series of useful tasks such as the transport of large microobjects[18] and cells[19] in different media.

However, some challenges still remain to be resolved such as high toxicity of the fuel, which are the optimal dimensions or how to improve high efficiency of microjets. A reduction of the concentration of peroxide fuel would permit the manipulation of living cells for long periods of time, whereas faster microbots would imply higher propulsion power to carry heavy loads.

The speed of the catalytic microrobots can be controlled by changing the temperature of the solution[20] where they move, reducing at the same time the concentration of fuel needed and performing ultrafast motion.

Rolled-up nanotechnology method allows also the fabrication of microbots (from trapezoid patterns) with sharp ends.[21] By incorporating magnetic layers, we can power, propel and guide those tubular microrobots towards soft biomaterials and drill into tissues. Magnetic microbots avoid the use of toxic fuels.[22] Also, the hollow structure of microtubes can allocate motile cells, which can be employed as hybrid bio-micro-robots.[23]

Here we review some of our recent developments on tubular microbots, from self-propulsion at low Reynolds numbers, external guidance for delivery of microobjects, collective motion and alternatives to fabricate biocompatible smart microbots. The working conditions, speed, size, and strengths and weaknesses of these self-propelled motors are summarized in Table 1. Our motivation is to extrapolate automated macroscale functions at the nanoscale, which is very relevant for emergent fields such as nano- engineering and nanorobotics. Some macroscopic tasks are extremely difficult at small scales in fluids, since it is hard to deliver power and control device functions accurately. Particularly, the first step towards realization of biomedical microbots consists on the design and fabrication of small tools that can mimic the functions of larger tools utilized in surgery.[24] We envision that these nano-/microrobots would have potential applications in biomedicine such as drug delivery and minimally invasive surgery, as well as sensing in microscale. The use of micro and nanorobots for advanced tasks such as drilling of cells or tissues will be described. First, we will present catalytic nanorobots that are remotely guided towards fixed cells where they are embedded. Next, we present our advances towards the combination of catalytic micromotors with biological fluids and cells. In order to get self-propulsion with reconstituted blood samples, thermal activation was needed.

The second part of this chapter, we present non-catalytic microbots: conical microdrillers magnetically actuated and micro-bio-robots that use propulsion energy from motile cells. This chapter finalizes with a summary and our vision on future perspectives on the field of microbots.

Table 1. Summary of the working environment, speed, size, strengths and weakness for different types of self-propelled motors presented in this chapter

	Environment	Speed	Size	Strengths	Weaknesses
Catalytic nano-robots	High concentrated H_2O_2 + surfactants aqueous solution	Up to 180 $\mu m/s$	300 nm in diameter, 6 – 10 μm in length	Nanometer scale in diameter, High efficiency, remote control, etc.	Toxic fuel solution is required.
Catalytic micro-robots	H_2O_2 + surfactants aqueous solution	From 100 $\mu m/s$ up to 10 mm/s depending on temperature	5 – 10 μm in diameter, 50 – 100 μm in length	micrometer scale in diameter, high efficiency, remote control, versatile functionalizations, etc.	Toxic fuel solution is required.
Magnetic micro-drillers	Aqueous solution	~ 5 $\mu m/s$ locomotion in planar direction and up to 1200 rpm in rotation	2 – 10 μm in diameter, 50 μm in length	Remote control, fuel free, etc.	Lack of directionality
Micro-bio-robots	Physiological solution	~10 $\mu m/s$	5 - 8 μm in diameter, 50 μm in length	biocompatibililty, remote control, fuel free, etc.	Slow, low power output

2   Microrobots Performing Complex Tasks

We present here different types of tubular microrobots fabricated by the "rolled-up nanotechnology" based on strain engineering of thin solid films. Detailed fabrication methods have been reported previously elsewhere[14,15] and brief description of each type of micro-robot will be given in following subsections. Their magnetic properties were exploited for the remote magnetic guidance using permanent magnets or more sophisticated computerized setups for the close-loop 2D control.[18,25,26,27]

2.1   Catalytic Nanojets Drilling into Fixed Cells

One class of tools in surgery is the sharp surgical instruments that are widely utilized for making incisions. Some of these surgical instruments are enabled by electromagnetic motors on the macroscale, but it is challenging to harness the energy in a tether-free manner required to perform drilling at smaller size scales.

Catalytic platinum constituted micro- and nano- structures accelerate the decomposition of hydrogen peroxide and enable the self-propulsion of micro- and nanomotors, pumping of fluids, and transport of colloidal particles and cell.

Such miniaturized and remote-controlled nanotools may have high potential for *in vivo* applications in the near future in the circulatory, the urinary and the central nervous systems.[28,29] However, to fabricate cost-effective and operative MIS devices, scientists need fabrication techniques that enable mass production of complex shaped three dimensional structures, and diverse types of materials. In this context, rolled-up nanotechnology – previously envisioned for nanodriller applications[30] – meets the above describe requirements.

We fabricated catalytic tubes with diameters in the sub-micrometer range and investigated control over their catalytic motion. By using molecular beam epitaxy (MBE), thin films of InGaAs/GaAs were deposited on sacrificial AlAs layers and bulk GaAs substrate, and a thin catalytic Pt film sputtered on top.[31] By rolling up those nanomembranes, we fabricated catalytic nanotubes with diameters approximately 20 times smaller than previously reported rolled-up microjets and half the size of the recently designed nanojets. Consequently, we reported the smallest man-made catalytic jet engines.[31]

The catalytic nanojets are powered by the decomposition of H_2O_2 into molecular oxygen, which accumulates in microcavity and eventually gets released from one end of the nanotube as visible micro-bubbles (Fig. 1). Fig. 1 illustrates the motion and trajectories of InGaAs/GaAs/Cr/Pt (3/3/1/1 nm) nanojets immersed in hydrogen peroxide fuel. These results demonstrate that micro-bubble driven catalytic nanojets can indeed overcome Brownian diffusion as well as the high viscous forces of the fluid at low Reynolds numbers. We can intentionally design their structure so that nanotubes present a sharp tip, clearly seen in Fig. 1 and Fig 3A. The release of bubbles from these rolled-up structures is asymmetric due to asymmetrically rolled-up layers. Thus, motion of catalytic nanojets in corkscrew trajectories can be designed by considering strain gradients and orientation of hybrid catalytic/heteroepitaxial thin films (Fig 1A).

Subsequently, we exploited the corkscrew propulsion (Fig 1A) of the nanojets to drill into biomaterials such as those constituting Hela cells, which are an immortal cell line derived from cervical cancer. It should be noted that we utilized paraformaldehyde to fix the cells prior to the drilling experiments for two reasons:

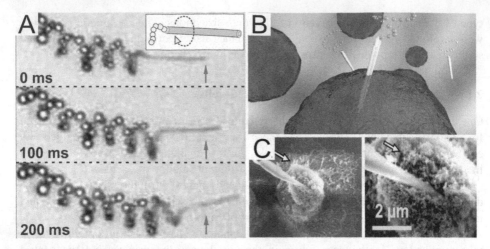

Fig. 1. (A) Corkscrew-like motion of a rolled-up microjet. Schematic (B) and SEM images (C) of a rolled-up microjet embedded into cells. Adapted from ref. [21].

(a) we wanted to remove the influence of any chemically induced deformation of the cell during drilling in the H_2O_2 fuel, (b) These fixed cells represent a cross-linked version of a realistic cellular biomaterial, so we rationalized that if the nanotools could generate enough force to drill into fixed cells, they would likely have more than enough force to drill into un-cross-linked cells. The type of motion needed for drilling is clearly shown in Fig. 1 by optical microscope sequences of an individual nanojet, which self-propels in a screw-like motion during 200 ms at a rotational frequency of 10 Hz (fuel composition: 20 % v/v H_2O_2, 10 % v/v surfactant). Straight arrows in the images indicate the linear displacement of the nanojet during the studied time. The inset of Fig. 1A depicts a schematic of the rotation of the nanojet during translation. The schematic image in Fig. 1B displays nanotools which self-propel and embeds itself into a fixed Hela cells. Once the cellular boundary is reached, the nanotools stick to it and start drilling into the cellular biomaterial over several minutes (Fig. 1C).

Although the fuel employed for self-propulsion is still toxic to sustain viable mammalian cellular functions, alternative mechanisms of powered motion and working conditions foresee the use of this concept in diverse applications such as biomedical engineering, biosensing and biophysics. While hydrogen peroxide may be acceptable for applications in nano-manufacturing and nanorobotics, biocompatible fuels need to be developed for live-cell applications. Nonetheless, due to the reduced dimensions but yet the high propulsion power, our results suggest strategies of using shape, size and asymmetry of catalytic nanostructures as tools to realize mechanized functions at the nanoscale.

2.2   Thermal Activation of Catalytic Microrobots in Blood Samples

The vision that intelligent robots could one day navigate along blood streams still remains unrevealed. We demonstrated that rolled-up micro-robots can self-propel in 10x diluted samples of blood at 37°C, i.e. physiological temperatures (see Figure 2A).[32] This work was by no means claiming that the current state of research in self-propelled microrobots enables their motion in blood streams *in vivo*. We aimed to proof that the high viscosity of blood at room temperature will not allow the motion of catalytic microrobots, but the propulsion in warmer solutions of diluted blood samples is still possible.

Microbots were fabricated by rolled-up nanotechnology consisting of Fe/Pt nanomembranes (6/1 nm) deposited onto photoresist squared patterns of 50 μm^2. An optical microscopy image from a microjet moving at 37°C in a 1 mL suspension of red blood cells (RBC) 10x diluted (5 x 10^5 RBC μL^{-1}) is illustrated in Figure 2B, reaching an average speed of 47 μm s^{-1}. We performed temperature cycles of five minutes at 25 and 37°C respectively, proving that catalytic microjets are able to self-propel in high complex media, i.e. RBC 10x diluted with a solution of serum at 10 wt.% (see Figure 2C), which may correspond to a 10 times dilution of a real blood sample, resulting in a final concentration of $5x10^5$ RBC μL^{-1}. In this approach, the viscosity of the solution is decreased and the activity of micromotors is enhanced by increasing the temperature of the solution. Previously, we reported the superfast motion of the same type of microrobots at physiological temperatures in aqueous peroxide solutions.[20]

As illustrated in Figure 2C, the microjets were not able to move significantly at 25°C. Due to the increase of viscosity after the addition of RBC in serum media, the absolute average speed at all temperatures decreased, reaching values from 15 to 25 μm s^{-1}. To demonstrate the effect of viscosity on the activation of the microjets, we monitored this parameter during the temperature cycles. As expected, and observed on Figure 2, viscosity is decreased when physiological temperatures are applied (37°C), going to its initial value once room temperature is recovered. Finally, we evaluated the microjets' performance in a microfluidic chip testing the same conditions for RBC 10x diluted in a 10 wt.% serum media (Figures 2D and 2E). Using the microfluidic approach, it is possible to reduce the consumption of reagents (i.e. blood sample) which is beneficial for scarce and high valuable samples. Figure 2E illustrates a microjet moving in the microfluidic channel when the temperature was kept at 37°C. The high viscosity and the presence of RBC did not prevent the microjets from moving from one reservoir to the other. We observed the continuous motion of microjets in the microfluidic chip over periods of 5 minutes. This experiment demonstrates the proof-of-principle that microjets could be eventually employed for the active motion in lab-on-chip devices for, e.g. biosensing, isolation of components, separation of different particles on chip, etc.

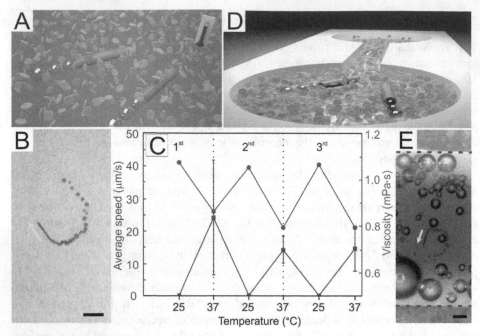

Fig. 2. Catalytic microrobots (microjets) in blood samples. (A) Schematic representation of behaviour of microjet engines in reconstituted blood samples. Microjets can self-propel if 10x diluted blood is kept at physiological temperature of 37°C. (B) Snapshot of a microjet moving in red blood cells (RBC) 10x diluted at 37°C. Scale bar:50 µm. (C) Average speed of Fe/Pt microjets warming up the RBC 10x diluted in 10 wt.% serum from 25°C to 37°C in three consecutive thermal cycles. (D) Sketch of the motion of microjets moving together with diluted blood in a microfluidic chip. (E) Optical microscope image of a microjet engine moving against flow in a microfluidic channel with RBC x10 times diluted in 10 wt.% serum at 37°C. Scale bar: 50 µm. Adapted from ref. [32].

2.3   Fuel Free Micro-drillers into Tissues Ex-vivo

To circumvent the limited applications of toxic fuel in vivo, an attractive approach relies on the fabrication of "fuel-free" tools, e.g. those powered by external magnetic fields. Recently, the enzymatically-triggered and tetherless thermobiochemical actuation of miniaturized grippers and tools, magnetically guided into liver tissues, was demonstrated.[33]

With the same rolled-up technology, we fabricated tubular Ti/Cr/Fe micro-drillers containing sharp tips (Fig 3A) that can be applied for mechanical drilling operations of porcine liver tissue *ex vivo* (Fig. 3B).[22] An external rotational magnetic field is used to remotely locate and actuate the micro-drillers in a solution with a viscosity comparable to that of biological fluids (e.g., blood). Changes in the frequency of the rotating magnetic field results in the switching of the rotational orientation of the micro-driller from a horizontal to a vertical one, which lifts the tubes and makes them suitable for drilling purposes. When microtools are place on hard planar surfaces (e.g. glass or silicon) and re-orient to the upright rotation, they are able to "walk" towards the center of the rotational magnetic field.

To demonstrate the drilling operation (Fig. 3B), a pig liver section was placed at the centre of the magnetic field in a petri dish containing microtools in the working solution (soap-water, 50 % (v/v)). The angular frequency was increased to 1150 rpm at which the microtools switched their orientation from horizontal into vertical one. Thereafter, the microtools were guided to the desired locations and started the drilling operation from tens of minutes to few hours. It was observed that the microtools retain upright orientation and the initial rotation frequency (~1150 rpm) immediately after reaching the tissue, but significantly slowed down in rotation frequency to few hundreds rpm (~400 rpm) after several minutes standing on the tissue, indicating the increase in friction during drilling into the tissue. However, that is not the case for microtools rotating on rigid glass surface, where they rotate at frequencies similar to the applied external rotation field (~1150 rpm). For a typical drilling operation, a depth of *ca.* 25 μm into the tissue could be achieved as judged via the length of the part of the driller remaining outside(Fig. 3C, upper panel). A strong permanent magnet was used to pull such embedded drillers out of the tissue leaving holes of approximatey same diameters in the liver section (Fig. 3C lower panel) indicating some tissues could be taken during the collecting.

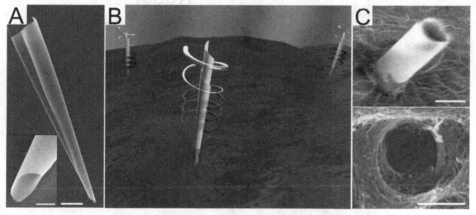

Fig. 3. Magnetic microdrillers. A) SEM images of a sharp microtube. B) Schematic of the motion and drilling of microtubes into tissues *ex-vivo* using rotating magnetic field. C) **(upper panel)** SEM image of a microdriller embedded into the pig liver section after drilling; **(lower panel)** SEM image showing the drilled hole in the pig liver section after extracting the microdriller by a strong permanent magnet (500 mT). Scale bar: 1μm in (A), 5μm in (B) and (C). Adapted from ref. [22]. Scale bar 5 μm in A and C, 1 μm in inset A.

2.4   Tubular Micro-bio-robots Using Motile Cells

We employed conical hollow tubes to capture motile bovine sperm cells and develop hybrid micro-biorobots (Figure 4).[23] Hybrid devices that harness biological energy for the propulsion of man-made microdevices have been one approach to develop novel microrobots. Magnetotactic bacteria[34] that offer remote control by applying external magnetic fields, as well as other microorganisms integrated in nano- and microdevices have been shown to be useful for creating micro-bio-robots. The motion control mechanisms of these hybrid devices vary from geometrical asymmetry,[35] chemotaxis,[36] magnetotaxis to electrokinetic and optical control[37], just to mention a few. All of the above systems use motile microorganisms as propulsion force and aim at developing robotic tools for applications in life sciences. Our group developed a new bio-hybrid microswimmer[23] by capturing bovine sperm cells inside ferromagnetic microtubes that use the motile cells as driving force. These micro-bio-robots can be remotely controlled by an external magnetic field. The combination of a biological power source and a microdevice is a compelling approach to the development of new microrobotic devices with fascinating future applications. This microswimmer might have significant impact in the biomedical field, in particular *in-vivo* fertilization methods where the controlled transport of a single sperm cell to the egg cell location is desired.

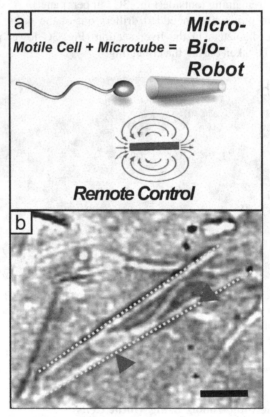

Fig. 4. (a) The micro-bio-robot is comprised of a motile sperm cell trapped inside a rolled-up microtube with incorporated magnetic layer for remote control. (b) Optical image of sperm cell (yellow shade) trapped inside a conically shaped Ti/Fe microtube (highlighted with yellow dots). Blue arrow points at sperm head, red arrow points at sperm flagella. Scale bar: 10 μm. Adapted from ref. [23].

3   Conclusion

We demonstrated that tubular microrobots can be employed for self-propulsion at small scales with high power output, versatility in applications towards bio-relevant issues. Catalytic bubble-propelled microrobots continuously developed and nowadays, the wireless control, transport of particles, capture of cells, and motion in microfluidic chips with biological samples has become a reality.

In addition, while using the same fabrication technology, the fabrication of 3D ferromagnetic microdrillers with sharp tips for drilling operation of soft biomaterials was reported. We presented magnetic control, drilling and guidance of fuel free microtools toward tissue samples ex-vivo. We also demonstrated that such incision can be performed in a fluid with viscosity similar to blood, which is ideal for future use in the field of microrobotics for minimally invasive surgery. The advantage of the tubular structure of the microtools is that the hollow structure might be utilized in the future for filling up with drug carrying gels for site directed drill and delivery systems, e.g., cholesterol degrading enzymes for clearing the arterial blockages and plaque removal nanorobots for minimal invasive surgery.

But also, tubular structures can capture single motile cells which are used as motor force of the hybrid motors. We expect that tubular micro-and nano-architectures will develop more in the near future for more complex tasks.

Acknowledgement. We thank D. H. Gracias for generating ideas and contribution in experimental work, C. Deneke for growing of the MBE samples, S. M. Harazim, A. N. Ananth for help with experiments. S.S., V. M. and W.X. thank the Volkswagen Foundation (86 362). S.S. thanks the European Research Council (ERC) for Starting Grant "Lab-in-a-tube and Nanorobotics biosensors", n° 311529. S.S. thanks R. Träger for schematics in Fig.1,2 and 3 and cover. S.S. and L.S thank DFG (Grant SA 2525/1-1) for financial support.

References

1. Kay, E.R., Leigh, D.A., Zerbetto, F.: Synthetic Molecular Motors and Mechanical Machines. Angew. Chem. Int. Ed. 46, 72–191 (2007)
2. Vallee, R.B., Hook, P.: Molecular Motors: A Magnificent Machine. Nature 421, 701–702 (2003)
3. Sen, A., Ibele, M., Hong, Y., Velegol, D.: Chemo and Phototactic Nano/Microbots. Faraday Discuss. 143, 15–27 (2009)
4. Paxton, W.F., Kistler, K.C., Olmeda, C.C., Sen, A., St. Angelo, S.K., Cao, Y., Mal-louk, T.E., Lammert, P.E., Crespi, V.H.: Catalytic Nanomotors: Autonomous Movement of Striped Nanorods. J. Am. Chem. Soc. 126, 13424–13431 (2004)
5. Mallouk, T.E., Sen, A.: Powering Nanorobots. Sci. Am. 300, 72–77 (2009)
6. Ozin, G.A., Manners, I., Fournier-Bidoz, S., Arsenault, A.: Dream Nanomachines. Adv. Mater. 17, 3011–3018 (2005)
7. Wang, J.: Can Man-Made Nanomachines Compete with Nature Biomotors? ACS Nano 3, 4 (2009)

8. Mirkovic, T., Zacharia, N.S., Scholes, G.D., Ozin, G.A.: Fuel for Thought: Chemically Powered Nanomotors Out-Swim Nature's Flagellated Bacteria. ACS Nano 4, 1782–1789 (2010)
9. Wang, J., Manesh, K.M.: Motion Control at the Nanoscale. Small 6, 338–345 (2010)
10. Laocharoensuk, R., Burdick, J., Wang, J.: Carbon-Nanotuble-Induced Acceleration of Catalytic Nanomotors. ACS Nano 2, 1069–1075 (2008)
11. Kline, T.R., Paxton, W.F., Mallouk, T.E., Sen, A.: Catalytic Nanomotors: Remote-Controlled Autonomous Movement of Striped Metallic Nanorods. Angew. Chem. Int. Ed. 44, 744–746 (2005)
12. Pumera, M.: Electrochemically powered self-propelled electrophoretic nanosubmarines. Nanoscale 2, 1643–1649 (2010)
13. Solovev, A.A., Mei, Y.F., Urena, E.B., Huang, G., Schmidt, O.G.: Catalytic Microtubular Jet Engines Self-Propelled by Accumulated Gas Bubbles. Small 5, 1688–1692 (2009)
14. Mei, Y.F., Huang, G.S., Solovev, A.A., Urena, E.B., Monch, I., Ding, F., Reindl, T., Fu, R.K.Y., Chu, P.K., Schmidt, O.G.: Versatile Approach for Integrative and Functionalized Tubes by Strain Engineering of Nanomembranes on Polymers. Adv. Mater. 20, 4085–4090 (2008)
15. Mei, Y.F., Solovev, A.A., Sanchez, S., Schmidt, O.G.: Rolled-Up Nanotech on Polymers: from Basic Perception to Self-Propelled Catalytic Microengines. Chem. Soc. Rev. 40, 2109–2119 (2011)
16. Harazim, S.M., Xi, W., Schmidt, C.K., Sanchez, S., Schmidt, O.G.: Fabrication and Applications of Large Arrays of Multifunctional Rolled-Up SiO/SiO2 Microtubes. J. Mater. Chem. 22, 2878–2884 (2012)
17. Manesh, K.M., Cardona, M., Yuan, R., Clark, M., Kagan, D., Balasubramanian, S., Wang, J.: Template-Assisted Fabrication of Salt-Independent Catalytic Tubular Microengines. ACS Nano 4, 1799–1804 (2010)
18. Solovev, A.A., Sanchez, S., Pumera, M., Mei, Y.F., Schmidt, O.G.: Magnetic Control of Tubular Catalytic Microbots for the Transport, Assembly, and Delivery of Micro-objects. Adv. Mater. 20, 2430–2435 (2010)
19. Sanchez, S., Solovev, A.A., Schulze, S., Schmidt, O.G.: Controlled Manipulation of Multiple Cells Using Catalytic Microbots. Chem. Commun. 47, 698–700 (2011)
20. Sanchez, S., Ananth, A.N., Fomin, V.M., Viehrig, M., Schmidt, O.G.: Superfast Motion of Catalytic Microjet Engines at Physiological Temperature. J. Am. Chem. Soc. 133, 14860–14863 (2011)
21. Solovev, A.A., Xi, W., Gracias, D.H., Harazim, S.M., Deneke, C., Sanchez, S., Schmidt, O.G.: Self-Propelled Nanotools. ACS Nano 6, 1751–1756 (2012)
22. Xi, W., Solovev, A.A., Ananth, A.N., Gracias, D.H., Sanchez, S., Schmidt, O.G.: Rolled-Up Magnetic Microdrillers: Towards Remotely Controlled Minimally Invasive Surgery. Nanoscale 5, 1294–1297 (2013)
23. Magdanz, V., Sanchez, S., Schmidt, O.G.: A Sperm Driven Micro-Bio-Robot. Adv. Mat. 25(45), 6581–6588 (2013)
24. Bassik, N., Brafman, A., Zarafshar, A.M., Jamal, M., Luvsanjav, D., Selaru, F.M., Gracias, D.H.: Enzymatically Triggered Actuation of Miniaturized Tools. J. Am. Chem. Soc. 132, 16314–16317 (2010)
25. Zhao, G., Sanchez, S., Schmidt, O.G., Pumera, M.: Micromotors with Built-In Compasses. Chem. Commun. 48, 10090–10092 (2012)
26. Sanchez, S., Solovev, A.A., Harazim, S.M., Schmidt, O.G.: Microbots Swimming in the Flowing Streams of Microfluidic Channels. J. Am. Chem. Soc. 133, 701–703 (2011)

27. Khalil, I.S.M., Magdanz, V., Sanchez, S., Schmidt, O.G., Abelmann, L., Misra, S.: Magnetic Control of Potential Microrobotic Drug Delivery Systems: Nanoparticles, Magnetotactic Bacteria and Self-Propelled Microjets. In: 35th Annual International Conference of the IEEE Engineering in Medicine and Biology Society, pp. 5299–5302. IEEE Press, New York (2013)
28. Nelson, B.J., Kaliakatsos, I.K., Abbott, J.J.: Microrobots for Minimally Invasive Medicine. Annu. Rev. Biomed. Eng. 12, 55–85 (2010)
29. Peyer, K.E., Zhang, L., Nelson, B.J.: Bio-Inspired Magnetic Swimming Microrobots for Biomedical Applications. Nanoscale 5, 1259–1272 (2013)
30. Schmidt, O.G., Eberl, K.: Nanotechnology: Thin Solid Films Roll Up into Nanotubes. Nature 410, 168 (2001)
31. Sanchez, S., Solovev, A.A., Harazim, S.M., Deneke, C., Mei, Y.F., Schmidt, O.G.: The Smallest Man-Made Jet Engine. Chem. Rec. 11, 367–370 (2011)
32. Soler, L., Martínez-Cisneros, C., Swiersy, A., Sánchez, S., Schmidt, O.G.: Thermal activation of catalytic microjets in blood samples using microfluidic chips. Lab Chip 13, 4299–4303 (2013)
33. Leong, T.G., Randall, C.L., Benson, B.R., Bassik, N., Stern, G.M., Gracias, D.H.: Tetherless thermobiochemically actuated microgrippers. Proc. Natl. Acad. Sci. U. S. A. 106, 703–708 (2009)
34. Martel, S., Tremblay, C.C., Ngakeng, S., Langlois, G.: Controlled manipulation and actuation of micro-objects with magnetotactic bacteria. Appl. Phys. Lett. 89, 233904 (2006)
35. Angelani, L., Di Leonardi, R., Ruocco, G.: Self-Starting Micromotors in a Bacterial Bath. Phys. Rev. Lett. 102, 048104 (2009)
36. Kim, D., Liu, A., Diller, E., Sitti, M.: Chemotactic steering of bacteria propelled microbeads. Biomed. Microdevices 14, 1009–1017 (2012)
37. Steager, E.B., Sakar, M.S., Kim, D.H., Kumar, V., Pappas, G.J., Kim, M.J.: Electrokinetic and optical control of bacterial microrobots. J. Micromech. Microeng. 21, 035001 (2011)

Addressing of Micro-robot Teams and Non-contact Micro-manipulation

Eric Diller, Zhou Ye, Joshua Giltinan, and Metin Sitti

Department of Mechanical Engineering
Carnegie Mellon University
5000 Forbes Ave
Pittsburgh, PA 15213, USA
sitti@cmu.edu

Abstract. This manuscript presents two methods for the addressable control of multiple magnetic microrobots. Such methods could be valued for microrobot applications requiring high speed parallel operation. The first uses multiple magnetic materials to enable selective magnetic disabling while the second allows for independent magnetic forces to be applied to a set of magnetic micro-robots moving in three dimensions. As an application of untethered magnetic microrobots, we also present a non-contact manipulation method for micron scale objects using a locally induced rotational fluid flow field. The micro-manipulator is rotated by an external magnetic field in a viscous fluid to generate a rotational flow field, which moves the objects in the flow region by fluidic drag. Due to its untethered and non-contact operation, this micro-manipulation method could be used to quickly move fragile micro-objects in inaccessible or enclosed spaces such as in lab-on-a-chip devices. In addition to introducing the operation and capability of these fabrication and control methods, we discuss the implications of scaling these systems to smaller scales for comparison with other microrobotics actuation and control techniques.

1 Introduction

The control of multiple micro-robots could have a major impact to enable parallel and distributed operation for manipulation, distributed sensing and other tasks in inaccessible micro-scale spaces [1]. Micro-robots controlled and powered using magnetic fields have gained use recently because they can be controlled remotely using relatively large magnetic forces and torques, can be a bio-compatible actuation method, and they can often be fabricated simply [2, 3]. However, methods to control teams of magnetic microrobots have been limited in the literature to crawling on planar surfaces [4, 5], and have issues of scalability. Here we present two new methods for the independent control of magnetic micro-robot teams. The first method uses composites of two magnetic materials to achieve on/off magnetization of individual micro-robots. By controlling the states of each agent, control over the set is achieved. This method is scalable to large arrays of micro-robots. The second addressing method achieves multi-robot control for

I. Paprotny and S. Bergbreiter (Eds.): Small-Scale Robotics 2013, LNAI 8336, pp. 28–38, 2014.

micro-robots levitating in a liquid medium for three-dimensional (3D) motion. Such 3D independent control has not been shown before, and is here accomplished by designing each micro-robot to respond uniquely to the same input magnetic fields.

Micro-object manipulation has wide potential applications in microfluidics, biological and colloidal science, lab-on-a-chip systems, and micro-assembly for systems such as micro-optical electro mechanical systems. Various techniques have been developed to achieve micro-manipulation in different backgrounds. These techniques can be categorized into two groups, contact-based manipulation and non-contact-based manipulation. Micro-grippers are major member of the first type [6, 7], while externally controlled bacteria also fall into this group [8, 9]. Many untethered micro-robots can also achieve manipulation tasks using mechanisms in the first category [10]. The second category includes optical tweezers [11], magnetic tweezers [12], dielectrophoresis [13], electrophoresis [14], optical arrays, and microfluidic devices, including use of micro-pumps/valves or magnetophoresis. Non-contact manipulation has also been implemented with several untethered micro-manipulators[15, 16].

We present a non-contact micro-object manipulation method using a locally induced rotational fluid flow field created by an untethered spherical magnetic micro-manipulator. The magnetic micro-manipulator is rotated in a viscous fluid by an externally generated magnetic field to create a rotational flow, which propels micro-objects in the flow region. One single spherical micro-manipulator is used to handle one object at a time. Automated manipulation of micro-beads is implemented based on visual feedback, with a very precise object position error of less than 20% of the object size.

2 Multi-robot Control Using Disabling Magnetism

2.1 Disabling by Magnetic Hysteresis

Remotely and selectively turning on and off the magnetization of many micro-scale magnetic actuators could be a great enabling feature in fields such as microrobotics and microfluidics. We have developed an array of addressable sub-millimeter micro-robots made from a composite material whose net magnetic moment can be selectively turned on or off by application of a large magnetic field pulse [17, 18]. The material is made from a mixture of micron-scale neodymium-iron-boron and ferrite particles, and can be formed into arbitrary actuator shapes using a simple molding procedure. To achieve many-state magnetic control of a number of microrobotic actuators, we require a number of magnetic materials with different hysteresis characteristics [19]. The magnetic coercivity and remanence (retained magnetization value when the applied field H is reduced to zero) for commonly-used materials vary over several orders of magnitude. These materials cover a wide range of hysteresis values, from NdFeB and SmCo, which are permanent under all but the largest applied fields, to iron, which exhibits almost no hysteresis. For comparison, the magnetic fields applied to actuate magnetic microactuators are typically smaller than 12 kA/m, which is only strong enough to remagnetize iron. Thus, the magnetic states of SmCo, NdFeB, ferrite and alnico can be preserved when driving an actuator. This can be used to independently control the magnetization of each material, even when they share the same workspace. By applying a pulse in the desired direction greater than the coercivity field (H_c) of a particular material, an independent magnetization state of each magnet material can be achieved instantly.

In general it is difficult to demagnetize a single magnet by applying a single demagnetizing field because the slope of the hysteresis loop (i.e. the magnetic permeability) near the demagnetized state is very steep. Thus, such a demagnetization process must be very precise to accurately demagnetize a magnet. While steadily decreasing AC fields can be used to demagnetize a magnetic material, this method does not allow for addressable demagnetization because it will disable all magnets in the workspace. This motivates the use of a magnetic composite to enable novel untethered addressable magnetic disabling.

We employ a different demagnetization procedure to achieve a more precise demagnetization by employing two materials, both operating near saturation where the permeability is relatively low. In this method, an applied switching field H_{pulse} can be applied to switch only one material's (ferrite in this example) magnetization without affecting the second material (NdFeB). This switching allows the device to be switched between on and off states as the magnetic moments add in the on state or cancel in the off state. While the internal field of the magnet at any point will not be zero, the net field outside the magnet will be nearly zero, resulting in negligible net magnetic actuation forces and torques.

When fields are applied below the NdFeB coercivity, the NdFeB acts as a permanent magnet, biasing the device magnetization, as shown in the H-m loop of Fig. 1 for H_{pulse} up to about 240 kA/m. This behavior is similar to that of electropermanent magnets, demonstrated for modular robotics applications [20]. Traversing the hysteresis loop, the device begins in the off state at point A, where motion actuation fields, indicated by the 12 kA/m range, only magnetize the device to about 0.08 μAm^2, resulting in minimal motion actuation. To turn the device on, a 240 kA/m pulse is applied in the forward direction, bringing the device to point B. After the pulse, the device returns to point C, in the on state. Here, motion actuation fields vary the device moment between about 1.7 and 1.8 μAm^2. To turn the device off, a pulse in the backward direction is applied, traversing point D, and returning to the off state at point A at the conclusion of the pulse. For small motion actuation fields in the lateral direction, the device is expected to show even lower permeability in the on or off state due to the shape anisotropy induced during the molding process.

2.2 Addressable Control Results

A six-electromagnetic-coil system is used to generate the magnetic field required to actuate and control the magnetic micro-manipulator. The system consists of three pairs of air-core independent electromagnetic coils, aligned to the faces of a cube approximately 8.2 cm on a side. A maximum magnetic field of 5 mT can be generated in the workspace, with 6% uniformity over a 30 mm space. The presented

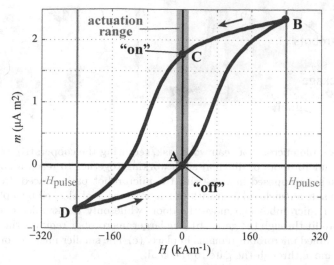

Fig. 1. The $H - m$ hysteresis loop of a composite microrobot made from ferrite and NdFeB. A 240 kA/m, field switches the ferrite magnetization while leaving the NdFeB unaffected, resulting in a vertically-biased loop intersecting the origin, showing clear on and off states.

disabling method for mobile microrobots can be used to selectively disable multiple microrobots. Based on its orientation when the pulse is applied (and independent of its position), each microrobot will be enabled or disabled. To achieve address-ability, the orientation of each must be controlled, which can be done using mag-netic field gradients, as described in [17]. Four and six microrobots are moved using stick-slip motion on a glass slide surface in a viscous oil environment. The viscous fluid environment is provided here to increase the fluid drag to retain microrobot orientation during the pulse. The experimental workspace is placed inside the coil system, allowing for both stick-slip motion on the 2D surface using small magnetic fields less than 12 kA/m and magnetic state changes by a larger field pulse. Inde-pendent addressing of the on and off states of each microrobot is accomplished by H_{pulse} applied in-plane.

An example of two micro-robots accomplishing a teamwork task is shown in Figure 2. Here, the task requires both a strong micro-robot to move the door and a small micro-robot to reach the goal through the small gate. Such a task is representative of complex tasks in micro-assembly and micro-transport which could benefit from heterogeneous groups of micro-robots.

The addressable control of multiple microrobots using magnetic disabling could be scaled to smaller or larger length scales. The operation depends criti-cally on the coercivity properties of magnetic materials, which can be preserved for most materials at particle sizes down to tens of micrometers. At smaller scales, the magnetic coercivity tends to decrease drastically. The fabrication of microrobots at smaller scales would require new methods, but it is expected that electroplating, polymer molding or sputtering of magnetic materials could

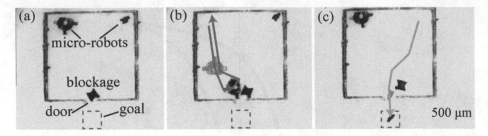

Fig. 2. Addressable microrobot teamwork task, requiring the cooperative contribution of two mobile microrobots of different sizes working together to reach a goal. Frames each show two superimposed images, with the microrobot paths traced. (a) Both microbots lie inside an enclosed area with the door to the goal blocked by a plastic door. Only the larger microrobot can move the door, while only the smaller microrobot is small enough to fit through the gate. (b) The larger microrobot removes the door while the smaller disabled microrobot remains in place. (c) The smaller microrobot is enabled and is free to move through the gate to the goal.

be used to reduce the microrobot size by perhaps one order of magnitude. The method could be used at larger scales without difficulty, and fabrication methods would become even more straighforward as traditional machining could be used.

3   Addressable Micro-robots in 3D

As an addressing method for magnetic microrobots which is capable of operation in 3D, we present the first microrobot addressing method for 3D motion control, which works by magnetic gradient pulling [21]. This method allows for completely independent and uncoupled net forces to be applied to each microrobot and thus allows for feedback position control of several microrobots in 3D. The magnetic force F exerted on a microrobot with magnetic moment m in a magnetic field B, assuming no electric current is flowing in the workspace, is given by

$$F = (m \cdot \nabla) B. \tag{1}$$

The application of different magnetic forces to each microrobot is accomplished by controlling the unique viscous drag and magnetic torque on each unique microrobot when placed in a rotating magnetic field. The different rotational responses of the geometrically and magnetically distinct designs result in each microrobot assuming a unique orientation in space, assuming a unique m. Thus, if each microrobot possesses a different magnetic strength or rotational fluid drag coefficient, arbitrary forces could be exerted on each independently and simultaneously using magnetic field gradients, when averaged over one short field cycle. The lag of a permanent magnet microrobot in a spinning magnetic field is determined by the balance between the applied magnetic torque and the drag torque. Fluid flow at the micro-scale is dominated by viscous forces as opposed to inertial forces, with typical Reynolds numbers of 0.01 or lower for the microrobot size scales around or smaller than 1 mm.

3.1   Addressable Microrobot Design

Various microrobots were fabricated to study the effect of the magnetic strength of the robot and the viscous drag on the lag of the robot. For small lag, it can be approximated as being proportional to the rotational frequency, where the proportionality constant is a function of the magnetic strength and viscous drag of the robot. The robots studied are shown in Figure 3. Placed in the same rotating magnetic field, these microrobots assume different angles from 0 to 90 degrees.

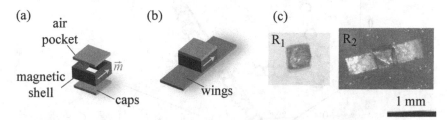

Fig. 3. Microrobots studied for 3D addressing. Robots R_1 and R_2 are given differing viscous drag coefficient but the same magnetic strength. Robots R_1 and R_2 are shown as molded and assembled.

3.2   3D Multi-robot Control Demonstration

An electromagnetic-coil system is used to generate the magnetic fields required to actuate and control the magnetic micro-robots. The system consists of eight coils which are aligned pointing to a common workspace center point with an approximate opening size of 12 cm. Using iron cores, a maximum magnetic field of about 20 mT can be generated in the workspace, with 6% uniformity over a 30 mm space. The feedback control methods are tested experimentally to prove their performance. Control using visual feedback is used to follow desired paths in 3D. Two micro-robots are moved along independent trajectories in 3D space in Figure 4. Here, the microrobots are controlled using the constant field rotation waveform, with a field rotation rate of 1.2 Hz.

Independent control of multiple microrobots using this method is expected to work across a wide range of microrobot size scales. The method relies on low-Re fluid drag relations, which stay valid down to size scales of nearly 1 μm. As the method could be implemented with permanent or non-permanent magnetic materials, there is not a barrier to scaling smaller from a magnetic actuation standpoint (although the efficiency of magnetic gradient pulling at small scales does degrade relative to other methods [22]). However, fabrication challenges to create neutrally-buoyant microrobots with well-defined 2D or 3D geometries may limit the practicality of this method below sizes of roughly 100 μm.

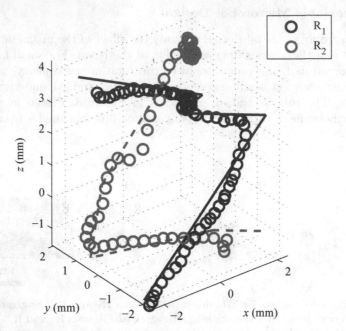

Fig. 4. Feedback control for two microrobots following the desired 3D paths in silicone oil with a viscosity of 0.052 Pa s. Circles show the tracked microrobot position every 1.0 s.

4   Non-contact Micro-manipulation

Teams of mobile microrobots can be used as mobile micro-manipulators for remote object transport and assembly tasks. Here we introduce similar microrobots which are used to achieve manipulation of objects in a non-contact manner by using the fluid flow generated around a spinning and translating micro-robot.

4.1   Fabrication of Spherical Manipulators

The magnetic micro-manipulators used in this study are fabricated via a soft-lithography-based micro-molding process, with slight differences from that in previous sections to accomodate the molding of spherical shapes. Small drops of solder are dropped into water to form the positive shapes for the spherical micro-manipulators, ranging from 10 μm to 1000 μm. The desired sizes are selected from the batch of solder spheres. These selected solder spheres are glued to a glass substrate using a UV curable epoxy (Loctite 3761) and a mold-making elastomer (PDMS, Dow Corning HS II RTV) is poured over to form a negative mold. A mixture of neodymium-iron-boron (NdFeB) particles (Magnequench MQP-15-7) suspended in polyurethane (TC-892, BJB Enterprises) matrix is poured into the negative rubber mold and allowed to cure into the final micro-manipulator shapes.

4.2   Experimental Demonstration

Experiments are carried out in a 30 mm x 30 mm x 1 mm open-top container filled with silicone oil with a kinematic viscosity of 0.052 Pa s. to create a low Re environment (Re<1). The spherical magnetic micro-manipulators used in all the experiments have a diameter of 360 μm, and the objects being manipulated are polystyrene beads with a diameter of 200 μm and a density of 1.05 g/cm.

The capabilities of rotational micro-manipulator locomotion and non-contact micro-object manipulation are shown in Figure 5, where a magnetic micro-manipulator transports a micro-bead on a glass substrate. An external magnetic field with strength of 3.5 mT rotating at a frequency of $f = 30$ Hz is applied to induce synchronous rotation with the applied field. In this experiment, the rotation axis angle is kept small (<10° from vertical), resulting in a translational manipulator speed of approximately 5 manipulator-diameters per second (about 900 μm/s).

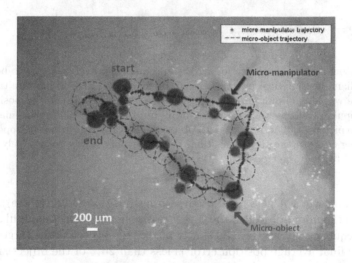

Fig. 5. Top-view optical microscope images of a 360-μm-diameter spherical magnetic micro-manipulator carrying a 200-μm-diameter polymer micro-bead along an arbitrary path on a glass substrate using rotational fluid field. Eight frames taken from video of the whole manipulation process, with an equal interval of 1.23 s between each of them, are overlaid to show the paths of the micro-manipulator and polymer bead. The entire duration is 9.87 s. The magnetic micro-manipulator rotates at 30 Hz and translates at a speed of approximately 2.5 manipulator-diameters per second (900 μm/s).

Automated course/fine manipulation of a 200 μm micro-object is also implemented based on visual feedback, with results illustrated in Figure 5. First, the micro-manipulator slowly rolls from the initial position to the standby position close to the micro-object (Figure 5 (ab)). Then it starts spinning at a high frequency ($f = 30$ Hz), picks up the micro-object (Figure 5 (c)) and carries it quickly towards the target position (Figure 5 (d)). The micro-manipulator stops

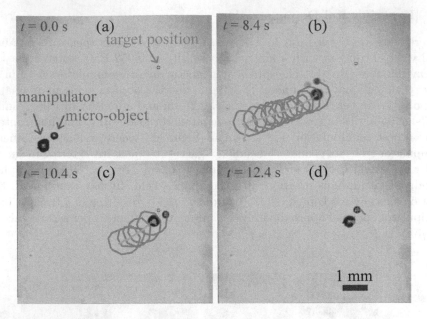

Fig. 6. Automated manipulation of a 200-μm-diameter polystyrene micro-bead using a 360-μm-diameter spherical magnetic micro-manipulator. Five frames are taken from video of the whole manipulation process at different periods. (a) Initial position. The micro-manipulator slowly rolls to the standby position. (b-c) The micro-manipulator fast spins to pick up and carry the micro-object to the target position. It stops spinning when getting close to the target position, then (d) spins slowly to precisely push the micro-object to the target position.

rapid spinning when it gets close to the target position (Figure 5 (e)). To precisely place the micro-object to the target position, the micro-manipulator spins at a low frequency ($f = 6$ Hz) and slowly pushes the object to the target (Figure 5 (f)). The final average position error is less than 20% of the object size.

4.3   Influence of Scale

The introduced manipulation mechanism relies purely on low-Re hydrodynamics, where Re is the dimensionless number characterizing the significance of inertial effects to viscous effects in fluidic systems. Since Re determines the hydrodynamic similarity between fluidic systems, the hydrodynamics in any two systems can be expected the same if these systems have similar Re's. Therefore, the manipulation mechanism should be applicable to situations where the low Re condition (Re<1) is satisfied, regardless of the absolute dimensions of the actual systems of robot or object. Although the geometric scale of the robot-object system has less direct impact on the hydrodynamics, it could have much significant influence on some other aspects. For example, when the size of the system decreases to a few microns or even sub microns, Brownian motion, which is generally negligible at larger scales, would change the object's motions significantly. In addition, if the length scale further reduces to molecular scales, the continuum hypothesis fails, and hence the hydrodynamics based on such hypothesis which our manipulation mechanism relies on. Therefore, whether or not the introduced mechanism would still be sufficient to perform manipulation tasks would need more rigorous investigation for such small scales.

5 Conclusions

We have presented two methods for the addressable control of multiple magnetic micro-robots in 2D and 3D. The first method has been shown to be scalable to larger groups of microrobots, while the second method has achieved multi-robot control in 3D fluid environments. Overcoming such limitations could allow for micro-robot teams to be used in high-impact parallel manipulation tasks. As an application for untethered micro-robots, we then propose using highly mobile rotating magnetic micro-robots to manipulate micro-objects via induced fluid flow at low Reynolds numbers. This method has promise to be a fast and precise non-contact object manipulation scheme for micro-fluidic environments. This method could benefit from multiple micro-robots working in parallel for high-speed team manipulation. These methods we demonstrated at the millimeter size scale, but work is progressing on reducing the microrobot size to as small as 5 μm for applications in microfluidics and bio-manipulation.

References

[1] Diller, E., Sitti, M.: Micro-scale mobile robotics. Foundations and Trends in Robotics 2(3), 143–259 (2013)
[2] Kummer, M., Abbott, J., Kratochvil, B., Borer, R., Sengul, A., Nelson, B.: OctoMag: An electromagnetic system for 5-DOF wireless micromanipulation. IEEE Transactions on Robotics 26(6), 1006–1017 (2010)
[3] Pawashe, C., Floyd, S., Sitti, M.: Modeling and experimental characterization of an untethered magnetic micro-robot. The International Journal of Robotics Research 28, 1077–1094 (2009)
[4] Diller, E., Floyd, S., Pawashe, C., Sitti, M.: Control of multiple heterogeneous magnetic microrobots in two dimensions on nonspecialized surfaces. IEEE Transactions on Robotics 28(1), 172–182 (2012)
[5] Diller, E., Pawashe, C., Floyd, S., Sitti, M.: Assembly and disassembly of magnetic mobile micro-robots towards deterministic 2-D reconfigurable micro-systems. The International Journal of Robotics Research 30(14), 1667–1680 (2011)
[6] Carrozza, M., Dario, P., Menciassi, A., Fenu, A.: Manipulating biological and mechanical micro-objects using LIGA-microfabricated end-effectors. In: IEEE International Conference on Robotics and Automation, vol. 2, pp. 1811–1816 (1998)
[7] Kim, D., Kim, B., Kang, H., Ju, B.: Development of a piezoelectric polymer-based sensorized microgripper for microassembly and micromanipulation. In: International Conference on Intelligent Robots and Systems, pp. 1864–1869 (October 2003)

[8] Martel, S., Tremblay, C., Ngakeng, S., Langlois, G.: Controlled manipulation and actuation of micro-objects with magnetotactic bacteria. Applied Physics Letters 89, 233904 (2006)

[9] Behkam, B., Sitti, M.: Bacterial flagella-based propulsion and on/off motion control of microscale objects. Applied Physics Letters 90, 023902 (2007)

[10] Tottori, S., Zhang, L., Qiu, F., Krawczyk, K.K., Franco-Obregón, A., Nelson, B.J.: Magnetic helical micromachines: fabrication, controlled swimming, and cargo transport. Advanced Materials 24(6), 811–816 (2012)

[11] Grier, D.G.: A revolution in optical manipulation. Nature 424(6950), 810–816 (2003)

[12] Yan, J., Skoko, D., Marko, J.: Near-field-magnetic-tweezer manipulation of single DNA molecules. Physical Review E 70(1), 1–5 (2004)

[13] Chiou, P.Y., Ohta, A.T., Wu, M.C.: Massively parallel manipulation of single cells and microparticles using optical images. Nature 436(7049), 370–372 (2005)

[14] Kremser, L., Blaas, D., Kenndler, E.: Capillary electrophoresis of biological particles: viruses, bacteria, and eukaryotic cells. Electrophoresis 25(14), 2282–2291 (2004)

[15] Petit, T., Zhang, L., Peyer, K.E., Kratochvil, B.E., Nelson, B.J.: Selective trapping and manipulation of microscale objects using mobile microvortices. Nano Letters 12(1), 156–160 (2012)

[16] Ye, Z., Diller, E., Sitti, M.: Micro-manipulation using rotational fluid flows induced by remote magnetic micro-manipulators. Journal of Applied Physics 112(6), 064912 (2012)

[17] Diller, E., Miyashita, S., Sitti, M.: Remotely addressable magnetic momposite micropumps. RSC Advances 2(9), 3850–3856 (2012)

[18] Miyashita, S., Diller, E., Sitti, M.: Two-dimensional magnetic micro-module reconfigurations based on inter-modular interactions. The International Journal of Robotics Research 32(5), 591–613 (2013)

[19] Diller, E., Miyashita, S., Sitti, M.: Magnetic hysteresis for multi-state addressable magnetic microrobotic control. In: International Conference on Intelligent Robots and Systems, pp. 2325–2331 (2012)

[20] Gilpin, K., Knaian, A., Rus, D.: Robot pebbles: One centimeter modules for programmable matter through self-disassembly. In: IEEE International Conference on Robotics and Automation, pp. 2485–2492 (2010)

[21] Diller, E., Giltinan, J., Sitti, M.: Independent control of multiple magnetic microrobots in three dimensions. The International Journal of Robotics Research 32(5), 614–631 (2013)

[22] Abbott, J., Nagy, Z., Beyeler, F., Nelson, B.: Robotics in the small, part I: Microbotics. Robotics & Automation Magazine 14(2), 92–103 (2007)

Progress Toward Mobility
in Microfabricated Millirobots

Sarah Bergbreiter[1], Aaron P. Gerratt[2], and Dana Vogtmann[1]

[1] University of Maryland, College Park, USA
sarahb@umd.edu
[2] École polytechnique fédérale de Lausanne, USA

Abstract. Research on mobile millirobots has been ongoing for the last 20 years, but the few robots that have walked have done so at slow speeds on smooth silicon wafers. However, ants can move at speeds approaching 40 body lengths/second on surfaces from picnic tables to front lawns. What challenges do we still need to tackle for millirobots to achieve this incredible mobility? This chapter presents some of the mechanisms that have been designed and fabricated to enable robot mobility at the insect size scale. These mechanisms utilize new microfabrication processes to incorporate materials with widely varying moduli and functionality for more complexity in smaller packages. Results include a 4 mm jumping mechanism that can be launched over 30 cm straight up, an actuated jumping mechanism used as a catapult, and preliminary leg designs for a walking/running millirobot.

1 Introduction

In biology, it is not uncommon to find impressive locomotion in small packages. Cockroaches can run at speeds up to 50 body lengths per second [1] and ants less than 5 mm long have been demonstrated running at speeds approaching 40 body lengths per second [2]. Insects like the flea and froghopper can jump to heights over 100x their own length [3].

Insects like these easily satisfy the size, speed, terrain, stability, payload, and robustness required for applications ranging from search and rescue to monitoring civil infrastructure. Previous research in mobile millirobots (defined as sub-centimeter mobile robots, often with microscale features) has never come close to achieving or explaining the remarkable mobility seen in insects.

This impressive mobility in insects is due in part to complex mechanisms packaged at small size scales. Jumping insects use a variety of materials and complex mechanisms to store energy for high power jumps [3]. Biologists have hypothesized that the reason insects can run at high speeds across complex surfaces without perceptibly adjusting leg timing or gait patterns is that feedback is minimal, and dynamic stability lies in the mechanics of the insect's legs instead [4–6].

These results from biomechanics have inspired numerous macroscale robot designs. Noh's flea-inspired robot uses a combination of novel materials and bio-inspired mechanisms to store and release energy for jumping [7]. Robots like

I. Paprotny and S. Bergbreiter (Eds.): Small-Scale Robotics 2013, LNAI 8336, pp. 39–52, 2014.
© Springer-Verlag Berlin Heidelberg 2014

RHex, Sprawl, and DASH utilize preflexes in which a feedforward motor pattern is combined with properly tuned compliance in the robots' legs [8–11]. While these robots are orders of magnitude larger than the insects that inspired them, they demonstrate that compliance, new materials, and mechanism design will be important in the design of jumping and running millirobots that can move across rough terrain.

In contrast, previous millirobots have generally relied on microfabrication and microelectromechanical systems (MEMS). Yeh fabricated the first articulated 2-DOF legs folded from polysilicon sheets [12], and Kladitis used a similar polysilicon process to flip up a thermal actuator that was also a leg [13]. Hollar attached 50 μm thick, 1 mm long silicon legs with polysilicon pin hinges that successfully demonstrated autonomous pushups [14, 15]. While this research represents impressive breakthroughs, none of the robots walked forward. One of many reasons for this lack of success is that microfabrication is often limited to the same materials used in integrated circuits - typically silicon, silicon dioxide, silicon nitride, polysilicon, and metals [16]. These materials are brittle, limited to strains of several percent, and have moduli of 10s to 100s of GPa.

Polymers added to these millirobot designs have enabled significantly more robust and successful locomotion, albeit at slightly larger sizes. Ebefors demonstrated a 15 x 5 mm^2 millirobot which walked forward at speeds up to 6 mm/s on a smooth silicon wafer, carried 30 times its own weight, and survived relatively rough handling [17]. One of the enabling features of this robot's success and robustness was the inclusion of thermally-activated polymer actuators in its legs [18]. Erdem used similar polymer-based thermal bimorphs on a 3 cm^2 robot that also walked [19]. A joint made of carbon fiber and polyimide resulted in an additional passive degree of freedom that enabled the first liftoff of a centimeter-scale flapping robot [20]. Clearly, it is promising to consider the benefits of adding compliant materials to millirobot design.

A key challenge in millirobotics is the addition of new materials to the currently existing microfabrication toolbox for manufacturing complex mechanisms and substantially improving locomotion. Poly(dimethylsiloxane) (PDMS) is a compliant material that can undergo elastic strains in excess of 100% and has a Young's modulus of 1.8 MPa [21], which is very similar to that of resilin, a biological material seen in insect wings and legs. PDMS is most commonly used to quickly and easily fabricate small and clear channels for fluid flow in microfluidics and bioMEMS [22]. However, it has rarely been used for its mechanical properties, primarily because of the lack of adequate fabrication processes. Parylene has been used to replace silicon springs given its modulus of 1 GPa [23], but this is still a relatively stiff material. Polyimide has been used for its thermal properties in the microrobot legs mentioned above [17] but also has a high modulus over 1 GPa.

In order to realize the benefits of low modulus materials in millirobots, this work presents a microfabrication process to incorporate compliant elastomer structures in-plane with traditional silicon microelectromechanical systems (Section 2). By incorporating new materials, elastomer springs are applied as compact energy storage mechanisms for small jumping robots and actuators are integrated to cat-apult projectiles off-chip (Section 3). Finally, preliminary leg designs for walking and running millirobots that also utilize this process are shown (Section 4).

2   New Materials and Fabrication

As described in the introduction, larger scale robots and insects owe part of their locomotion success to the use of diverse materials and mechanisms. For millirobots, feature sizes on the order of microns to tens of microns will be important along with a wide variety of material mechanical properties. Smart composite microstructures (SCM) come close to satisfying these constraints, but generally result in larger-scale features more appropriate for centimeter-sized robots [24]. Microfabrication provides the required feature sizes, but a common criticism for MEMS in millirobotics is the lack of material diversity.

To incorporate tailored compliance into millirobot legs for jumping and run-ning, elastomeric materials are embedded directly into a silicon MEMS process. Two processes are described here; the first molds elastomer components through the full thickness of a silicon wafer, while the second starts with a silicon-on-insulator (SOI) wafer and molds elastomer features through the top device layer. This process enables smaller feature sizes (2 – 10 μm) and easier integration with microfabricated actuators.

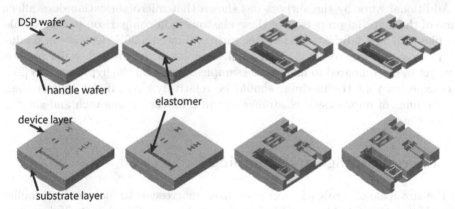

Fig. 1. The two fabrication processes developed to create elastomer-based mechanisms in microfabrication. The top process uses a through-wafer etch of a double side polished (DSP) wafer (typically 300–500 μm thick). The bottom process uses an SOI wafer (with a device layer 20–80 μm thick). Modified from [25].

The basic fabrication process has been described in detail in previous publi-cations by the authors, including [25], and is shown in Fig. 1. In both processes, a layer of silicon is etched using deep reactive ion etching (DRIE), resulting in trenches that are then re-filled with elastomer. In the first process, a double-side polished (DSP) wafer (typically 300 – 500 μm thick) is bonded to a handle wafer and etched. The second process etches trenches into the device layer (typically 20-80 μm thick) of a silicon on insulator (SOI) wafer instead.

Dow Corning Sylgard 184 is the most common elastomer used, although the process has been demonstrated with other silicones as well. This is a two part elastomer and is mixed thoroughly in a 10:1 ratio of the base to the curing agent before being poured over the surface of the wafer. To ensure that the elastomer completely refills the etched trenches, the wafer is put into a vacuum and held at 1 Torr for 10 minutes. The elastomer is then cured at 90 °C for two hours.

Once the elastomer is cured, the excess is removed from the surface of the wafer by running a razor blade across the surface of the wafer. This leaves some residual pieces of elastomer, so a one minute rinse in a 3:1 mixture of n-methylpyrrolidone and tetrabutylammonium fluoride is performed [26]. This planarizes the elastomer to the top of the wafer.

The new surface is planar enough that a new layer of photoresist can be deposited and patterned on the front side of the wafer. This creates a second mask which is used to perform a second DRIE through the thickness of the DSP wafer or device layer of the SOI wafer. This etch results in silicon features that are patterned around the elastomer features.

The final step in the process is to release the devices. In the case of the through wafer process, this release is from the handle wafer and accomplished by soaking the wafers in acetone and performing a brief etch in 6:1 buffered hydrofluoric acid (BHF) to remove the silicon dioxide mask. In the SOI process, this is accomplished using a backside DRIE etch and a timed etch in 6:1 BHF.

Additional work by the authors has shown that microfabrication does affect some of the material properties of these elastomers in comparison to macroscale samples [27]. SOI fabricated elastomer springs showed a 20% decrease in modulus, in large part due to the 6:1 BHF etch. Samples in the through-wafer process have yet to be compared to macroscale samples. Based on the hypotheses in [27], it is expected that the modulus should be relatively close, if not greater than the modulus of unprocessed elastomer due to a longer plasma etch and shorter release time.

3   Energy Storage and Jumping

As the size scale of a robot is decreased from macroscale to mesoscale to milliscale, challenges related to mobility increase as the size of a robot itself decreases relative to the objects in the environment around it. As a result, jumping becomes an attractive mobility method for millirobots [28]. A background on the mechanics of jumping millirobots is provided in [25, 29]. It is important to note that stored energy needs to be released in a relatively short time (milliseconds) given the millimeter-scale length of the robot's leg, and drag can have significant impact on the jumping height of small robots. While stored energy needs to be released quickly, it is important to note that it can be stored slowly which makes this task manageable with existing microactuation.

3.1 Jumping Mechanism

In order to examine jumping locomotion at the microscale, a jumping mechanism was fabricated and first presented in [29]. The robot mechanism is essentially two rigid masses that are connected by a series of elastomer springs, as shown in Fig. 2. In this case, the robot "leg" is in the center of the device and the robot "body" is the u-shape around the outside. When an external force is applied, the structure is compressed and the springs are strained, storing potential energy. When the external force is removed, the potential energy is released. The force in the springs works to increase the kinetic energy of the body.

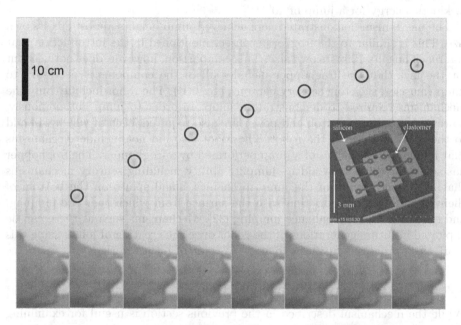

Fig. 2. Screenshots from a video showing the takeoff of the jumping mechanism. Due to the small size of the jumping mechanism relative to the scale of the jump, the mechanism is difficult to make out, but it has been circled in each frame to show the position. An SEM of the jumping mechanism is inset. Modified from [25].

By compressing the robot with tweezers and therefore tensioning the elastomer springs, the mechanism was repeatedly launched and reached a maximum height of 32 cm. A series of 18 jumps reached an average height of 19.4 cm with a standard deviation of 7.2 cm. Failed launches, defined as jumps that reach heights less than 2 cm, were not included in these statistics. The jumping performance varied greatly because of the method of launching the robot. It was not uncommon for the robot to hit the tweezers during or immediately after takeoff, which dramatically affected the jump performance. The height was determined by launching the robot in front of 1 cm grid paper, and video of several jumps was taken with a Casio Exilim EX-F1 camera at 300 frames per second. Screenshots from these videos are shown side-by-side in Fig. 2. The maximum jump height of 32 cm was 80x the robot's own height, and the same robot was used repeatedly, demonstrating the robustness of the process and final mechanism.

The jump that reached 32 cm had an initial velocity of 3 m/s, or initial kinetic energy of 36 μJ. This number was calculated by measuring the distance traveled between two frames of the video. This corresponds to a Reynolds number of 726 ($\rho_{air} = 1.2$ kg/m^3, $v = 3$ m/s, $L = 0.004$ m, and $\mu_{air} = 1.98 \times 10^{-5}$ kg/m/s). A Reynolds number this low means that drag should be considered. Additional losses due to spring viscosity, leg mass [30], rotation, and interaction with the tweezers during takeoff result in a transfer efficiency from stored potential energy to kinetic energy for a jump of 40%.

The mechanism demonstrated here achieved jump heights almost 80x its own size. This is similar to the froghopper insect mentioned in the introduction that can reach heights 100x its own size. This comparison, however, does not account for the fact that the froghopper includes all of the components necessary to jump (muscles, skeleton, sensory neurons, etc. [31]). The robot includes only the mechanisms required to demonstrate a jump. In order to jump autonomously, however, actuators, control, and power are also required, each of which will add to the mass and size of the robot. The robot also did not include mechanisms that mimic the mechanics of a jump performed by a froghopper. The froghopper has several features that aid its jumping ability including sensory mechanisms that aid in the timing of the jump mechanics, small spines on the bottom of their legs that increase friction with the surface from which they are jumping, and specialized legs to enhance jumping [32]. Mechanisms such as these can be exploited in future generations of the robot once the creation of joints using this fabrication process is further explored – a topic of future work.

3.2   Actuated Energy Storage and Release

While the mechanism described in the previous section is useful for examining the merits of jumping microrobots, it is interesting to consider how actuation can be added to such a system, as any system employed outside of a laboratory setting would require on-board actuation. To add actuation, the SOI process was used to define silicon features in the device layer of the SOI wafer.

A SEM image of a fabricated actuated mechanism is shown in Fig. 3, and more details are provided in [25]. The color in this image was added after it was captured for illustrative purposes. The frame is shown in dark gray with black patterning. The frame was used to transfer the force from the actuators to the springs. The springs, orange in color in Fig. 3, were anchored to the substrate, shown in black, at one end and were attached to the frame at the other end. There were two sets of actuators arranged around the frame; one set is blue and the other set is purple. Fig. 3 shows a mechanism with two springs, but mechanisms with eight springs (two sets of four) were also fabricated.

The actuators used in this work were thermal actuators, often referred to as chevron actuators because of their angled arms which resemble a chevron.

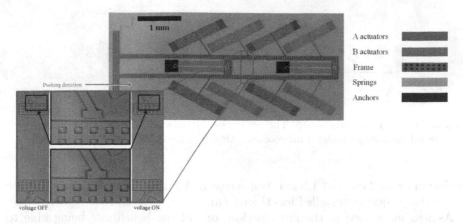

Fig. 3. A colored SEM image of the actuated mechanism. The blue and purple areas are the two sets of actuators. The orange areas are the springs. The black areas are the anchors for the springs and the patterned gray area is the frame. The zoomed area shows the interface between the actuator and the 'leg'. Modified from [25].

When a current is passed through the structure, the silicon heats up due to Joule heating, which results in expansion of the silicon. The chevron beams are at a slight angle, so the expansion leads to bending of the beams, which pushes the central beam forward to push a shuttle or leg. The actuation scheme was based on the design presented in the work by Maloney [33]. By operating the A and B sets of actuators in an alternating fashion, many small displacements of the frame were accumulated to result in a large displacement. The sequence began with both the A and B actuators on. Then the A actuators were turned off, allowing the B actuators to push. The A actuators were then turned back on, returning the actuators to the intermediate step with both actuators on. Finally the B actuators were turned off, allowing the A actuators to push. The process was repeated until the desired displacement was achieved.

The actuated mechanism was used to propel a projectile, as shown in Fig. 4. The system was cycled to strain the elastomer springs 45% and store 0.45 μJ. The projectile, an 0402 sized surface mount capacitor with a mass of 1.4 mg, was then placed directly in front of the frame. The actuators were then turned off, releasing the frame. The force from the energy stored in the springs worked to accelerate the frame, which in turn worked to accelerate the projectile. The projectile was 1 mm long, 0.5 mm wide, and 0.5 mm tall. In the best test, the projectile traveled more than 7 cm, so it traveled at least 70x the longest dimension of the projectile. The exact distance is not known because the projectile fell off the stage under the microscope after it traveled 7 cm. As with the tests performed on the jumping mechanism, not every test was successful. The projectile was 0.5 mm tall, but the frame of the actuated mechanism was only 0.02 mm tall, so during several tests the frame slid under the projectile. In a series of eight consecutive tests, five were successful and had projectile travel distances with an average and standard

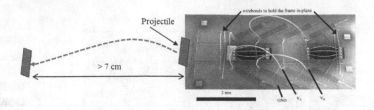

Fig. 4. The projectile test setup. The projectile was placed in front of the 'leg', which was placed on a stage under a microscope. Modified from [25].

deviation of 4.82 cm and 1.9 cm, respectively. A failed test was defined as one where the projectile travelled less than 1 cm.

As was mentioned in the introduction, one of the benefits of being able to integrate a material such as an elastomer is the ability to have repeatable actuation. Several tests were performed where the springs were strained to store energy and then released to allow the springs to return to their unstrained position. This cycle was then repeated to strain and release the springs a total of 10 times. This test was performed 17 times with one device at various operating voltages and frequencies without failure for a total of 170 strain/release cycles, demonstrating the robustness of the elastomer material. These tests strained the springs to several 10s of percent strain, but strains as high as 100% were demonstrated. The fastest speed demonstrated was 0.7 mm/s to achieve 45% strain. This was an average speed over the entire strain cycle as the speed decreased as the strain, and therefore force in the springs, increased.

4   Joints and Legs

Combined elastomer and silicon mechanisms can also be used to create distributed leg compliance and damping that will improve locomotion over relatively rough terrain. The ultimate goal of this work is to explore leg designs that can enable insect-like legged locomotion in millimeter-scale robots. As a first step toward that goal, elastomer joints were modeled and fabricated. At the millimeter-scale, integration of actuators, wiring, and mechanisms make it challenging to design and fabricate fully actuated legs. Instead, it will be important for millirobots to utilize passive degrees of freedom to improve locomotion. In addition, leg compliance has benefits beyond simplified fabrication as mentioned in the introduction. A simple dynamic model called the spring-loaded inverted pendulum (SLIP) shows that much of the speed and stability in insects is due to compliance in the legs [6, 4]. Most running animals regardless of their size and number of legs follow the same center of mass trajectory and ground reaction forces predicted by the SLIP model (although it is yet to be shown that this same phenomenon occurs at the millimeter-scale where inertia becomes less important – another future goal of this work).

4.1   Modeling Elastomer Joints

In order to design compliant legs, it is first necessary to understand the behavior of elastomer joints fabricated using the process in Section 2. Modeling of miniature compliant mechanisms is most commonly accomplished using methods such as the pseudo-rigid body (PRB) method, which replaces the compliant hinge with a pin joint and torsion spring [34]. This approach works well for small angular deflections and when axial lengthening is not signficiant. However, flexure joints made from soft materials like PDMS are very likely to have large angular deflections and significant axial extensions.

Finite element analysis (FEA) tends to be highly accurate, but time-consuming and computationally expensive compared to the PRB method. Previous work by the authors has shown that a 3-spring PRB model consisting of two torsional springs bracketing a single axial spring can accurately capture large deflections and axial extension without requiring the computational complexity of FEA [35].

More detail is provided in [35], but the basic spring constants for this model have been determined using geometric and material parameters, according to the following equations:

$$k_{\theta 3spr} = \frac{2EI}{l_{eff}} \tag{1}$$

$$k_{l3spr} = \frac{CAE}{l_{eff}} \tag{2}$$

where $k_{\theta 3spr}$ is the torsional spring constant, E is the elastic modulus of the joint material, I is the cross-sectional moment of inertia of the joint, l_{eff} is an effective hinge length, k_{l3spr} is the axial spring constant, C is a correction factor based on the geometry of any adhesion features in the joint, and A is the cross-sectional area of the joint.

In these equations, $k_{\theta 3spr}$ is similar to the single spring PRB model, distributed across two torsion springs, with the exception that length l is replaced with l_{eff}, an effective hinge length also based on the joint geometry. The linear spring constant k_{l3spr} depends on the cross-sectional area of the hinge and uses the equation for axial beam stiffness, again using l_{eff}. The correction factor, C, can also be included to account for reduced stiffness due to adhesion geometry.

4.2   Fabricated Joints

Using the SOI fabrication process described in Section 2, elastomer-based compliant joints have been fabricated and tested. As can be seen in Fig. 5, the joints can be manipulated in any direction. The joint shown in Fig. 5 was rotated in-plane 90° over 100 times without failure. Even when stressed, the hinge performed well and snapped back into position when released.

However, some of the joints designed in this initial phase suffered due to designed lengths that were only a few times longer than they were tall. These joints preferred to bend at the interface with silicon as opposed to bending consistently

Pushed left   As fabricated   Pushed right

Fig. 5. Preliminary leg hinges in SOI-based fabrication process

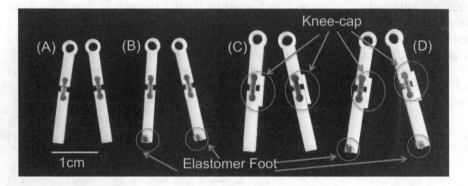

Fig. 6. Preliminary leg designs fabricated at approximately 10x scale

through the length of the joint as was expected. While the model described above has been experimentally and quantitatively validated at slightly larger scales, test structures have yet to be implemented at small scales to validate this 3-spring PRB model.

4.3   Preliminary Leg Designs

Leg designs using these joints have also been tested at larger scales (Fig. 6). These preliminary designs borrow many ideas from larger scale robots that are able to traverse relatively rough terrain at high speeds like RHex and iSprawl [10, 11]. In particular, the legs have one active joint that will be driven by a motor and one passive joint similar to that seen in Fig. 5. These larger-scale legs have also been designed while taking into account the constraints imposed by the fabrication process in Section 2. While RHex legs can rotate 360°, it is much more difficult to provide the same actuation in a microfabricated structure. This 360° rotation separates the 'step' phase in which the leg drives the robot forward, and the 'reset' phase in which the leg resets for its next step. This work assumes that such a motor is not available at the millimeter scale. Instead, a typical microactuator may only be able to rotate a leg back and forth through

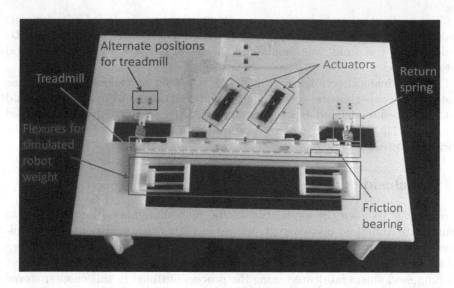

Fig. 7. The larger scale experimental setup with a leg, a randomized-terrain treadmill, and actuators

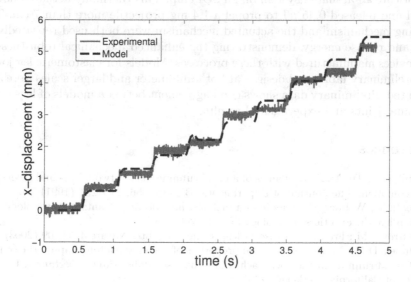

Fig. 8. Experimental and model treadmill displacement over 5 seconds. Leg type (B).

a prescribed arc, so new approaches are needed to separate these step and reset phases for a microfabricated leg.

The four different leg designs in Fig. 6 use a combination of asymmetries in the form of an elastomer 'foot' and a knee-cap to more clearly separate the step and reset phases in legged locomotion. The elastomer foot provides more grip during the step phase than the reset phase and the knee-cap locks the knee in place during the step phase. These asymmetries were included in a dynamic model of two legs in MSC ADAMS. Joints were modeled using the 3-spring PRB model and a 'treadmill' was simulated to measure the walking speed and total distance traveled.

The experimental setup shown in Fig. 7 was used to validate this dynamic model. Early experimental results for the Type (B) legs in Fig. 6 show a good match between model and experimental results (Fig. 8). Current work is focused on reproducing these same experimental results at the millimeter-scale using the fabrication process in Section 2.

5   Conclusions

This work has shown the application of the first microfabrication processes to incorporate compliant elastomer structures in-plane with traditional silicon microelectromechanical systems. By incorporating new materials, elastomer springs are applied as compact energy storage mechanisms for small jumping robots. A jumping mechanism fabricated using the process outlined in this chapter stored 100 μJ, 40% of which was transferred into kinetic energy of the mechanism resulting in jump heights as high as 32 cm. A similar fabrication process was used to fabricate an actuated system on an SOI chip. This thermally actuated device stored and released 0.45 μJ to propel a 1.4 mg projectile more than 7 cm. The jumping mechanism and the actuated mechanism were both used repeatedly to store and release energy, demonstrating the enhanced mechanical robustness of the devices manufactured with these processes. Models for elastomeric leg joints and preliminary fabricated designs at both millimeter and larger scales were also presented. Preliminary data shows good agreement between models of legs using elastomer joints and experimental results.

References

1. Full, R.J., Tu, M.S.: Mechanics of a rapid running insect: two-, four- and six-legged locomotion. The Journal of Experimental Biology 156, 215–231 (1991)
2. Seidl, T., Wehner, R.: Walking on inclines: how do desert ants monitor slope and step length. Frontiers in Zoology 5(1), 8 (2008)
3. Burrows, M.: Froghopper insects leap to new heights. Nature 424, 509 (2003)
4. Dudek, D.M.: Passive mechanical properties of the exoskeleton simplify the control of rapid running in the cockroach, Blaberus discoidalis. PhD dissertation, University of California, Berkeley (2006)
5. Kubow, T., Full, R.: The role of the mechanical system in control: A hypothesis of self-stabilization in hexapedal runners. Philosophical Transactions of the Royal Society London B 354, 849–862 (1999)
6. Spagna, J.C., Goldman, D.I., Lin, P.C., Koditschek, D.E., Full, R.J.: Distributed mechanical feedback in arthropods and robots simplifies control of rapid running on challenging terrain. Bioinspiration & Biomimetics 2, 9–18 (2007)
7. Noh, M., Kim, S.W., An, S., Koh, J.S., Cho, K.J.: Flea-inspired catapult mechanism for miniature jumping robots. IEEE Transactions on Robotics 28(5), 1007–1018 (2012)

8. Brown, I., Loeb, G.: A reductionist approach to creating and using neuromusculoskeletal models. In: Biomechanics and Neuro-Control of Posture and Movement, pp. 148–163. Springer, New York (2000)
9. Saranli, U., Buehler, M., Koditschek, D.E.: Design, modeling and preliminary control of a compliant hexapod robot. In: IEEE International Conference on Robotics and Automation, San Francisco, CA (2000)
10. Saranli, U., Buehler, M., Koditschek, D.E.: RHex: a simple and highly mobile hexapod robot. International Journal of Robotics Research 20(7), 616–631 (2001)
11. Cham, J.G., Bailey, S.A., Cutkosky, M.R.: Robust dynamic locomotion through feedforward-preflex interaction. In: ASME International Mechanical Engineering Congress and Expo (November 2000)
12. Yeh, R., Kruglick, E.J.J., Pister, K.S.J.: Surface-micromachined components for articulated microrobots. Journal of Microelectromechanical Systems 5(1), 10–17 (1996)
13. Kladitis, P.E., Bright, V.M.: Prototype microrobots for micro-positioning and micro-unmanned vehicles. Sensors and Actuators A: Physical 80(2), 132–137 (2000)
14. Hollar, S., Flynn, A.M., Bellew, C., Pister, K.S.J.: Solar powered 10 mg silicon robot. In: IEEE Micro Electro Mechanical Systems, pp. 706–711 (2003)
15. Hollar, S., Flynn, A.M., Bergbreiter, S., Pister, K.S.J.: Robot leg motion in a Planarized-SOI 2 poly process, Hilton Head, SC (June 2002)
16. Petersen, K.E.: Silicon as a mechanical material. Proceedings of the IEEE 70(5), 420–457 (1982)
17. Ebefors, T., Mattsson, J.U., Kalvesten, E., Stemme, G.: A walking silicon microrobot. In: International Conference on Solid-State Sensors, Actuators, and Microsystes (Transducers), Sendai, Japan, pp. 1202–1205 (June 1999)
18. Ebefors, T., Ulfstedt-Mattsson, J., Kalvesten, E., Stemme, G.: 3D micromachined devices based on polyimide joint technology. In: Conference on Devices and Process Technologies for MEMS and Microelectronics, SPIE, Gold Coast, Australia, vol. 3892, pp. 118–132 (October 1999)
19. Erdem, E.Y., Chen, Y.M., Mohebbi, M., Suh, J.W., Kovacs, G.T.A., Darling, R.B., Bohringer, K.F.: Thermally actuated omnidirectional walking microrobot. Journal of Microelectromechanical Systems 19(3), 433–442 (2010)
20. Wood, R.J.: The first takeoff of a biologically inspired at-scale robotic insect. IEEE Transactions on Robotics 24(2), 341–347 (2008)
21. Schneider, F., Fellner, T., Wilde, J., Wallrabe, U.: Mechanical properties of silicones for MEMS. Journal of Micromechanics and Microengineering 18(6), 065008 (2008)
22. Xia, Y., Whitesides, G.M.: Soft lithography. Annual Review of Materials Science 28(1), 153–184 (1998)
23. Suzuki, Y., Tai, Y.C.: Micromachined high-aspect-ratio parylene spring and its application to low-frequency accelerometers. Journal of Microelectromechanical Systems 15(5), 1364–1370 (2006)
24. Wood, R.J., Avadhanula, S., Sahai, R., Steltz, E., Fearing, R.S.: Microrobot design using fiber reinforced composites. Journal of Mechanical Design 130(5), 052304 (2008)
25. Gerratt, A.P., Bergbreiter, S.: Incorporating compliant elastomers for jumping locomotion in microrobots. Smart Materials and Structures 22(1), 014010 (2013)
26. Lee, J.N., Park, C., Whitesides, G.M.: Solvent compatibility of poly(dimethylsiloxane)-sased microfluidic devices. Analytical Chemistry 75(23), 6544–6554 (2003)

27. Gerratt, A.P., Penskiy, I., Bergbreiter, S.: In situ characterization of PDMS in SOI-MEMS. Journal of Micromechanics and Microengineering 23(4), 045003 (2013)
28. Bergbreiter, S.: Effective and efficient locomotion for millimeter-sized microrobots. In: IEEE/RSJ International Conference on Intelligent Robots and Systems, Nice, France, pp. 4030–4035 (September 2008)
29. Churaman, W.A., Gerratt, A.P., Bergbreiter, S.: First leaps toward jumping micro-robots. In: IEEE/RSJ International Conference on Intelligent Robots and Systems, San Francisco, CA, USA, pp. 1680–1686 (September 2011)
30. Burdick, J., Fiorini, P.: Minimalist jumping robots for celestial exploration. The International Journal of Robotics Research 22(78), 653–674 (2003)
31. Braunig, P., Burrows, M.: Neurons controlling jumping in froghopper insects. The Journal of Comparative Neurology 507(1), 1065–1075 (2008)
32. Burrows, M.: Morphology and action of the hind leg joints controlling jumping in froghopper insect. Journal of Experimental Biology 209, 4622–4637 (2006)
33. Maloney, J.M., Schreiber, D.S., DeVoe, D.L.: Large-force electrothermal linear mi-cromotors. Journal of Micromechanics and Microengineering 14, 226–234 (2004)
34. Howell, L.L.: Compliant Mechanisms. John Wiley & Sons, Inc., New York (2001)
35. Vogtmann, D., Gupta, S.K., Bergbreiter, S.: Characterization and modeling of elas-tomeric joints in miniature compliant mechanisms. ASME Journal of Mechanisms and Robotics 5(4), 41017 (2013)

From Nanohelices to Magnetically Actuated Microdrills: A Universal Platform for Some of the Smallest Untethered Microrobotic Systems for Low Reynolds Number and Biological Environments

Tian Qiu[1,2], John G. Gibbs[1], Debora Schamel[1,3], Andrew G. Mark[1], Udit Choudhury[1], and Peer Fischer[1,3]

[1] Max Planck Institute for Intelligent Systems, 70569 Stuttgart, Germany
[2] Institute of Bioengineering, Ecole Polytechnique Fédérale de Lausanne (EPFL), CH-1015 Lausanne, Switzerland
[3] Institut für Physikalische Chemie, Universität Stuttgart, Pfaffenwaldring 55, 70569 Stuttgart, Germany
fischer@is.mpg.de

Abstract. Building, powering, and operating structures that can navigate complex fluidic environments at the sub-mm scale are challenging. We discuss some of the limitations encountered when translating actuation mechanisms and design-concepts from the macro- to the micro-scale. The helical screw-propeller or drill is a particularly useful geometry at small scales and Reynolds numbers, and is one of the mechanisms employed by microorganisms to swim. The shape necessarily requires three-dimensional fabrication capabilities which become progressively more challenging for smaller sizes. Here, we report our work in building and operating these screw-propellers at different sizes. We cover the length scales from the sub 100 nm to drills that are a few hundred microns in length. We use a known physical deposition method to grow micron-sized magnetic propellers that we can transfer to solutions. We have recently succeeded in extending the fabrication scheme to grow nanohelices, and here we briefly review the technical advances that are needed to grow complex shaped nanoparticles. The microstructures can be actuated by a magnetic field and possible applications of the micro- and nanohelices are briefly discussed. We also present a system of polymeric micro-screws that can be produced by micro-injection molding and that can be wirelessly driven by an external rotating magnetic field through biological phantoms, such as agarose gels with speeds of ~200 μm/s. The molding technique faithfully reproduces features down to a few microns. These microdrills can serve as a model system to study minimally invasive surgical procedures, and they serve as an efficient propeller for wireless microrobots in complex fluids. The fabrication scheme may readily be extended to include medically approved polymers and polymeric drug carriers.

Keywords: low Reynolds number propulsion, microrobot, microdrill, microscrew, glancing angle deposition, micro molding, biological tissue.

I. Paprotny and S. Bergbreiter (Eds.): Small-Scale Robotics 2013, LNAI 8336, pp. 53–65, 2014.
© Springer-Verlag Berlin Heidelberg 2014

1   Introduction

Moving through fluid environments at the scale of microorganisms presents a different set of challenges compared to those encountered by macroscopic swimmers. Particularly at low Reynolds number (Re << 1), which indicates a Stokes regime of fluid flow with a dominance of viscous forces over inertial forces, it is known that a simple time reversible motion will not result in any net displacement of the swimmer [1]. Hence, asymmetric non-reciprocal actuation mechanisms are required at low Re. Microorganisms use two non-reciprocal propulsion mechanisms: the travelling wave beats of cilia and the helical rotation of flagella.

Mimicking a rotating flagellum requires a rotary motor and power source capable of producing sufficient torque to overcome the high viscous drag at low Reynolds number. One may consider the use of electromagnetic motors, which are ubiquitous in macro-scale robotics. However, electromagnetic motors require sizeable currents which preclude miniaturization. One of the smallest commercial electromagnetic motors is 6 mm long with a diameter of 1.9 mm [2]. This is too large for applications in micro-surgery. Piezoelectric rotary motors do not require large currents and piezoelectric elements can readily be obtained that have small linear dimensions (~250 μm), but they require relatively high input voltages ~28 V_{pp} [3]. If the motor is to be powered wirelessly using a battery, then this presents a problem, as thin film lithium ion batteries typically supply microampere currents at 1-3 V which corresponds to μW (for an area of ~20 mm^2). Similarly, microfuel cells would require at least 1 cm^2 area of each electrode to produce power in the range of mW [4]. There are therefore no simple compatible combinations of motor and onboard powering source for designing sub-millimeter micro-swimmers. Hence, we resort to external magnetic fields and torques. Magnetically actuated rotation can be achieved with micro- and nanostructures that contain a ferromagnetic material and that can be actuated by a homogenous magnetic (*i.e.* gradient-free) field. However, in order for a robot to be propelled by a gradient-free field at low Reynolds number an asymmetrical shape is essential. Propulsion in this regime has been achieved with rigid chiral nanostructures, *i.e.* solid helically-shaped micro-propellers [5, 6]. A helix breaks spatial symmetry in a manner that allows for low Reynolds number propulsion by coupling rotational and translational motion; as a helical micro-robot rotates about its long axis, this hydrodynamic coupling leads to propulsion along this axis. Various fabrication techniques exist for constructing helical micro-robots:

1. 20 nm – 300 nm: Micellar nanolithography and shadow deposition (Glancing angle deposition) on cooled substrates [7]
2. 300 nm – 10 μm: Glancing angle deposition [5]
3. 10 μm – 100 μm: Direct laser writing of helical structures [8]
4. 20 μm – 100 μm: Metallic thin film strain engineering techniques [6]
5. 100 μm – mm: Micro molding [this work]

Here, we review the fabrication scheme and principle of operation of the smallest magnetically actuated microbots that can currently be operated in liquids (1 and 2, above). We also present a low-cost bench-top micro-molding scheme that is able to produce polymeric magnetic micro-screws that can move in tissue phantoms (5, above). The choice of materials can thus be extended to medically approved polymers. Numerous applications can be proposed for the microbots at nano to micro length scales (1 and 2, above) especially in biological studies, *e.g.* as rheological probes to study the micro-rheology of complex biological media including cell membranes or as a carrier into the cell for genetic transfer; while the polymeric magnetic micro-screws (5, above) may serve as a micro-tool for biopsy in minimally invasive surgical procedures or as a vehicle for drug delivery.

2   Fabrication of Nanohelices and the Smallest Microbots

The fabrication technique that we focus on here is called glancing angle deposition (GLAD). With this method, a wide range of materials possessing many functionalities, such as ferromagnetism and electrical conductivity just to name a few, can be grown by physical vapor deposition, including magnetically-driven micro-robots [5]. The structures can be fabricated in large-numbers and with precisely defined geometries. This permits different length scales as well as different geometries to be realized that optimize the propulsive behavior at small scales [9].

GLAD is a physical vapor deposition technique [10-12]. A basic schematic is shown in Figure 1 (a). In a vacuum chamber at pressures of ~10^{-7} mbar, the source material is heated via electron beam bombardment until the material vaporizes. In the figure the vapor flux is for simplicity shown as a cone, but the vapor flux in general spreads with a broader angular distribution. Because of the low vacuum environment, the atoms impinge upon the substrate in a ballistic manner which is essential for taking advantage of the shadowing effect. Shadowing growth occurs when a surface onto which material is being deposited is tilted to a very oblique angle α, *i.e.* oblique angle deposition (OAD), which is typically $80° < \alpha < 87°$. Consider a substrate that is oriented at such an oblique angle. As the impinging vapor flux deposits on the substrate, if there is any surface roughness, *i.e.* raised portions of the substrate, the depositing material will preferentially accumulate on these raised features leaving the shadowed areas mostly free of material. If the substrate is perfectly flat, random nucleation sites will naturally form and serve as points of growth, but no regular ordering will be present in this case. If the substrate is intentionally seeded with well-ordered seeds, then the growth can be restricted to accumulate on the seed particles as will be discussed below.

The substrate, as shown in Figure 1 (a), is manipulated by two motors: the first motor, which is not shown in Figure 1, controls the vapor deposition angle α, which is defined as the angle between the flux and the substrate surface normal; the second motor controls the substrate rotation angle φ. No rotation of φ during deposition leads to arrays of nanorods tilted toward the plane of the substrate at an angle $\beta \neq \alpha$, whereas rapid rotation leads to growth of arrays of nanorods perpendicular to the surface. If the rotation is carefully controlled at intermediate speeds and in accordance with the rate of material deposition, an array of helices is produced. The helix pitch is inversely proportional to the rotation rate $d\varphi/dt$. It should be noted that the final morphologies are material-dependent and must therefore be tuned accordingly. For example, the substrate must be cooled significantly for materials with high adatom surface mobilities [7]. Multiple materials can be added subsequently for layered architectures or at the same time to produce various alloys. The three parameters that are used to characterize the morphology of an individual helix are presented in Figure 1 (b): P is the helix pitch, R is the helix major radius, and r is the helix minor, or wire radius. It should also be noted that there is a linear dependence upon the helix major radius and the helix pitch, but this is beyond the scope of the present article. Each parameter can be controlled to varying degrees as discussed below.

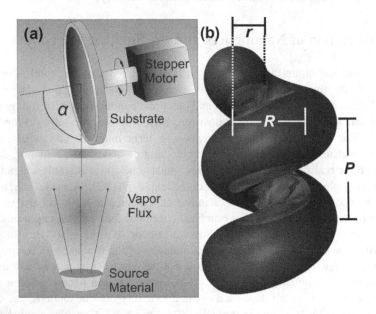

Fig. 1. (a) schematic of the GLAD process showing the source, vapor flux, and the orientation of the substrate; (b) schematic of an individual helix showing the helix minor radius r, helix major radius, R, and the helix pitch, P

The morphology of the helix can be controlled in the following ways: the size of the initial seeds onto which the helix is grown translates directly to the size of r and defines the range of possible P; the substrate rotation rate $d\varphi/dt$ defines P; and the material which is being deposited.

Although a wide range of sizes can be fabricated with GLAD onto a non-patterned surface, in order to have greater control over the final morphology of the helix, and to have uniformity between helices in the helix array, a properly seeded substrate should be used. Electron beam (ebeam) lithography [13] is excellent for designing seed patterns, but if rapid fabrication is required, one must use a rapid seeding technique because e-beam lithography is slow and can only cover small areas. Here we describe methods to rapidly seed the substrates with individual seed dimensions ranging from several microns to nanometers. In this manner we can reproducibly construct helical micro- and nano-robots over the entire wafer for over three orders of magnitude: from tens of nanometers to roughly ten microns.

The first technique requires patterning the wafer with an array of hexagonally close-packed spherical silica particles. This is accomplished using a Langmuir-Blodgett trough [14, 15]. As can be seen in Figure 2 (a), the ~ 400 nm SiO_2 beads are arranged in a close-packed arrangement on the surface of the substrate. It should be noted that defects in the monolayer crystal are present, but these are not important for our purposes here. For larger helices such as the 2-turn SiO_2 array shown in the cross-section SEM image in Figure 2 (b), this seeding approach is appropriate. These helices are then removed from the surface via sonication and suspended into a liquid, e.g. water. Individual helices which have been redeposited onto a new substrate for imaging are shown in the SEM image of Figure 2 (c) with a close up of a single helix in Figure 2 (d). This SiO_2 helix has a pitch P ~ 500 nm. The material is changed during growth and a magnetic material, Ni or Co, is sputtered onto the helices as shown in Figure 2 (c)-(d) [5]. The magnetic material is deposited at $\alpha = 0°$ and so coats only the top half. The helices are then magnetized in a manner that the magnetization direction is perpendicular to the helix axis. An alternative method is to incorporate the magnetic material into the structure during deposition as shown in Figure 2 (e) [16]. Clear contrast of the 150 nm of Co layer can be seen in the cross-section SEM image just above the silica beads.

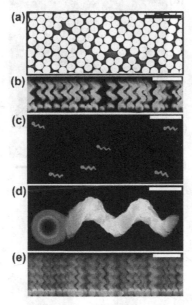

Fig. 2. (a) top-view SEM image of 400 nm SiO_2 beads on an Si(100) substrate; (b) side-view SEM of SiO_2 helices grown on the monolayer of beads; (c) individual helices removed from the array and redeposited onto another wafer; (d) a close-up of one of the same helices from (c); (e) a different 4-turn SiO_2 helix with a 150 nm layer of Co to add magnetic functionality as seen as the bright contrast just above the beads. Scale bar: (a), (b), (c), (d), and (e): 2 μm, 2 μm, 5 μm, 500 nm, and 2 μm respectively.

In order to produce nano-scale helices rapidly at the wafer-scale, we use the nanolithography process called block co-polymer micellar lithography (BCML) [17] to seed the substrate. BCML requires the self-assembly of polystyrene-b-poly[2-vinylpyridine (HAuCl$_4$)] diblock copolymer micelles. Plasma treatment removes the polymer and reduces the Au-salt leaving behind hexagonally-arranged Au nanodots as shown in the top-view SEM image of Figure 3(a). The BCML process allows for the separation between seeds and the size of the individual dots to be tuned. We have recently shown that using the combination of BCML and GLAD allows for the fabrication of nano-colloidal particles with tailored optical, electromagnetic, and mechanical properties [7]. An example oblique angle SEM image of Cu nanohelices is shown in Figure 3 (b) and a TEM of an individual Cu helix is shown in Figure 3 (c). We reduce the temperature of the substrate before and during deposition to aid the shadowing effect and to reduce adatom mobility which is a key to the fabrication of helices of this size. The minor radius, r, for these Cu helices is < 20 nm with the overall length ~ 150 nm. This advanced fabrication technique allows for the fabrication of magnetic nano-scale helices as well, although nano-propulsion with helices of this size has yet to be demonstrated.

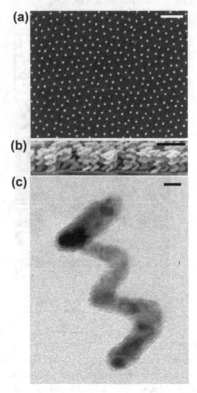

Fig. 3. (a) SEM top-view of a Si(100) wafer patterned via BCML nanolithography; (b) oblique-view SEM of 2.5-turn Cu helices grown on the BCML-patterned wafer; (c) TEM image of an individual Cu helix showing nanometer dimensions. Scale bar: (a), (b) 200 nm; and (c) 20 nm.

3   Magnetically Actuated Microrobots

The micron-sized helices described in the previous section can be diametrically magnetized by a strong magnet. A (weak) rotating homogeneous magnetic field B can now be used to couple to the magnetic moment of the helices and this causes a rotation of the helices around the long axis, which, due to the symmetry-breaking of their chiral structure, leads to rotation-translation-coupling and therefore to a forward propulsion [5]. In our setup we use 3-axis Helmholtz coils that can generate rotating magnetic fields in 3D of more than 100 Gauss from DC to higher than kilohertz frequency by integration of an active water-cooling system into the metal frame of the coil [16]. A drawing of the coil's frame and a picture of an assembled Helmholtz coil system in an inverted microscope are shown in Figure 4. The field direction and strength is controlled with a custom LabView program, which enables us to steer the micro-propellers in 3D on micron-length scale.

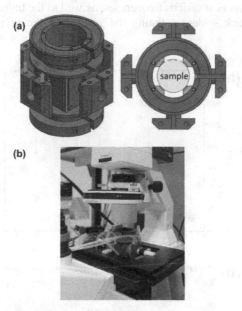

Fig. 4. (a) CAD drawing of the coil's frame (side and top view). The blue arrows indicate a water in-/outlet. (b) Image of the coil in an inverted microscope setup.

The propellers follow the magnetic field until the torque due to the applied magnetic field can no longer overcome the fluid's drag forces. This is called the step-out frequency. The translational speed of the helix depends linearly on the field's frequency up to the step-out frequency and can be described by the following general analytical equation which we have derived in [16]:

$$v = \varepsilon \, \Omega = \varepsilon \left(\omega - \frac{\kappa \; \sec^2\left[-\frac{t}{2C}\sqrt{\kappa} + \tan^{-1}\left[\frac{1}{\sqrt{\kappa}}\right]\right]}{C^2 \omega \left(1 + \left(-\frac{1}{C\omega} + \frac{\sqrt{\kappa}}{C\omega}\tan\left[-\frac{t}{2C}\sqrt{\kappa} + \tan^{-1}\left[\frac{1}{\sqrt{\kappa}}\right]\right]\right)^2\right)} \right) \tag{1}$$

where

$$\kappa = C^2\omega^2 - 1 \qquad (2)$$

and

$$C = \frac{X\,\eta}{M_{rem}\,B_0} \qquad (3)$$

Here ω is the rotational frequency of the field and Ω is that of the particle, and t denotes time. The propeller's remanent magnetization is M_{rem}, the strength of the applied magnetic field is B_0 and η is the viscosity. X is a size-invariant geometry factor that depends only on the shape of the particle (for a sphere $X=8\pi$). The propulsion efficiency ε determines the forward translational speed at a given frequency and is a direct measure of the strength of the translation-rotation coupling. Its upper limit is set by the screw's pitch [16, 18]. For our GLAD structures, ε has a value on the order of a few nm/rad, and we can achieve speeds of about 2.5 µm/s at a magnetic field strength of 50 Gauss [16]. Figure 5 shows the velocities of the swimmers at various frequencies, as well as the trajectory of one microbot controlled by a joystick – demonstrating the control of the propulsion trajectory on micron-length scales.

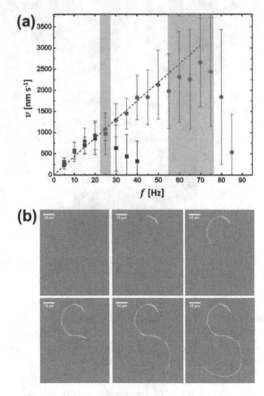

Fig. 5. (a) Propulsion speeds at various frequencies for magnetic field strengths of 20 and 50 Gauss (blue squares and red circles, respectively), with the step-out-frequencies indicated by the blue and red shaded areas (graph taken from Reference [16]), (b) trajectory of one actively driven micropropeller, demonstrating the control of the propulsion trajectory on micron-length scales

The small size of these microswimmers combined with the high accuracy with which they can be propelled make them promising candidates for manipulation of biological systems on small length-scales. They are usually made out of silica, which can easily be functionalized with various chemicals, such as enzymes or fluorescent dyes. We therefore expect a number of interesting applications such as remote sensing and local micro-manipulation to emerge in the near future.

4   Microdrills for Biological Environments

At larger length-scales we use a metal micro-screw as a template from which we mold polymeric microdrills. The template is prepared by electrical discharge machining (EDM). As shown in Figure 6, the drill is designed to have an outer diameter of 300 µm in order to fit inside a 23 gauge needle (nominal inner diameter 337 µm). Hardened steel is used as the template. The EDM process is time-consuming and is limited to conducting materials, which may not be suitable for medical applications. We have therefore developed a micro-molding process that uses a single EDM machined template from which polymer micro-screws may be batch-produced.

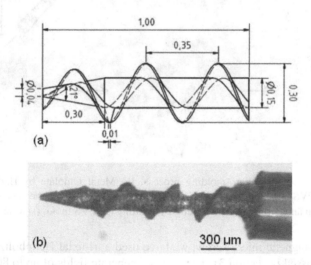

Fig. 6. The metal template for the micro-molding process. (a) Schematic drawing of the micro-screw design. Sizes are in mm. (b) Scanning electron microscope image of the micro-screw template manufactured by electro-discharge machining (EDM).

The micro-molding process consists of 6 steps, as illustrated in Figure 7. First, the metal template is manufactured by an EDM process (Institut für Mikrotechnik in Mainz, Germany (Figure 7 (a)). Then Polyvinyl siloxane (PVS) impression material (Art. No 4667, Coltene Whaledent, Switzerland) is mixed and the metal template is inserted (Figure 7 (b)). After 5 min curing, the metal template is removed (Figure 7 (c)). Cycloaliphatic Epoxide Resin (ERL-4221 Modified SPURR Embedding Kit,

SERVA Electrophoresis GmbH, Heidelberg, Germany) is then injected into the mold (Figure 7 (d)). After the epoxy is cured at 70°C for 3 hours, the PVS mold is cut and split, and the polymer micro-screw is released (Figure 7 (e)). Finally, a cylindrical NdFeB micro-magnet (200 μm in diameter and 400 μm in length) is attached to the end of the polymer (Figure 7 (f)). By this cheap and fast micro-molding process, the micro-structure of the metal template is precisely replicated.

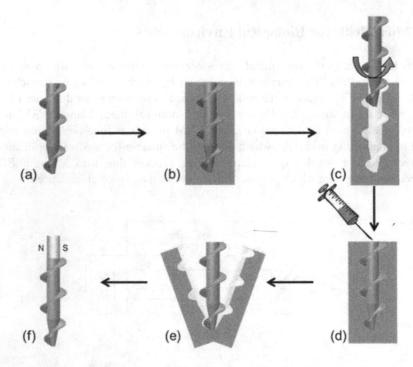

Fig. 7. Illustration of the micro-molding process. (a) Metal template by EDM. (b) Micro-molding using PVS. (c) After curing of the PVS mold, removal of the metal template. (d) Polymer injection and curing. (e) Unmolding by splitting the PVS mold. (f) Magnet attachment.

To test the magnetic micro-screw, we have used a tri-axial Helmholtz coil (similar to the one discussed in section 3). The coil can generate fields of up to 80 Gauss at up to 100 Hz (Figure 8). We use custom LabView software to control the amplitude and direction of the rotating magnetic field in 3D. To mimic the rheological properties of biological tissue we prepare various agarose gels for *in vitro* testing. Figure 9 shows a micro-screw that is propelled in agarose gels. We have tried propulsion in 0.1% wt-1% wt agarose gels, the latter requiring higher fields (up to 500 Gauss). The trajectory (see Figure 9 (d)) is defined in real time by a joystick. The average linear velocity reaches roughly 200 μm/s with a magnetic field rotating at 5 Hz.

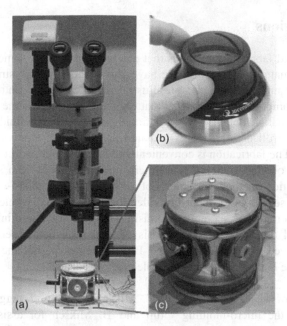

Fig. 8. Helmholtz coil setup to drive the micro-screw. (a) Tri-axis Helmholtz coil setup is used to drive and steer the micro-screw. Stereo-microscope (Leica MZ95 stereoscope with a Leica DFC 490 camera) is used to observe the movement. (b) 3-dimensional navigation of the micro-screw can be realized by turning the magnetic field with a joystick. (c) Enlarged picture of the tri-axial Helmholtz coil.

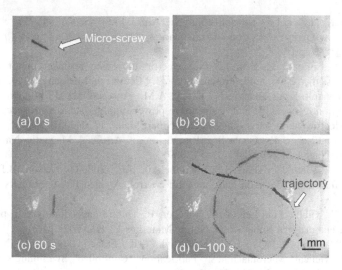

Fig. 9. Propulsion of the micro-screw in agarose gel. Snapshots after (a) 0 s, (b) 30 s, (c) 60 s, respectively, (d) the trajectory of the micro-screw from 0 to 100 s defined by the external magnetic field.

5   Conclusions

We have reviewed fabrication methods that can be used to grow some of the smallest magnetically actuated microrobots. We have shown how helical structures that are smaller than 100 nm can be made. We illustrated the excellent control that we have over material composition and shape that this fabrication scheme offers and the superior control that we obtain when actuating these structures in fluids. We have also demonstrated controlled magnetically-actuated propulsion of a polymer micro-screw in agarose gels. The fabrication is convenient and scalable and permits larger numbers of screws to be obtained quickly. The micro-molding process can serve as a cost-effective replication method for microbot propellers. The micro-screws have the potential to be used as an efficient propeller for self-powered wireless microbots in fluids. The systems may serve as promising micro-tools for minimally invasive therapeutics, and the fabrication scheme is general such that it permits the use of medically approved polymeric materials. Both schemes permit the use of surface chemistries or the loading with suitable molecules and drugs.

Acknowledgments. The authors thank C. Miksch for helpful suggestions and for assistance with the micro-molding setup and B. Miksch for assistance with the LabView program. This work was supported by the European Research Council under the ERC Grant agreement Chiral MicroBots (278213).

References

1. Purcell, E.M.: Life at Low Reynolds-Number. Am. J. Phys. 45, 3–11 (1977)
2. http://www.micromo.com/datasheets/BrushlessDCmotors/ 0206_B_DFF.pdf
3. Watson, B., Friend, J., Yeo, L.: Piezoelectric ultrasonic resonant motor with stator diameter less than 250 μm: the Proteus motor. J. Micromech. Microeng. 19, 022001 (2009)
4. Cook-Chennault, K.A., Thambi, N., Sastry, A.M.: Powering MEMS portable devices - a review of non-regenerative and regenerative power supply systems with special emphasis on piezoelectric energy harvesting systems. Smart Mater. 17, 043001 (2008)
5. Ghosh, A., Fischer, P.: Controlled Propulsion of Artificial Magnetic Nanostructured Propellers. Nano Letters 9, 2243–2245 (2009)
6. Zhang, L., Abbott, J.J., Dong, L.X., Kratochvil, B.E., Bell, D., Nelson, B.J.: Artificial bacterial flagella: Fabrication and magnetic control. Applied Physics Letters 94, 064107 (2009)
7. Mark, A.G., Gibbs, J.G., Lee, T.-C., Fischer, P.: Hybrid nanocolloids with programmed 3D-shape and material composition. Nature Materials 12, 802–807 (2013)
8. Tottori, S., Zhang, L., Qiu, F.M., Krawczyk, K.K., Franco-Obregon, A., Nelson, B.J.: Magnetic Helical Micromachines: Fabrication, Controlled Swimming, and Cargo Transport. Adv. Mater. 24, 811–816 (2012)
9. Keaveny, E.E., Walker, S.W., Shelley, M.J.: Optimization of Chiral Structures for Microscale Propulsion. Nano Letters 13, 531–537 (2013)

10. Hawkeye, M.M., Brett, M.J.: Glancing angle deposition: Fabrication, properties, and applications of micro- and nanostructured thin films. Journal of Vacuum Science & Technology A: Vacuum, Surfaces, and Films 25, 1317–1335 (2007)
11. Robbie, K., Brett, M.J.: Sculptured thin films and glancing angle deposition: Growth mechanics and applications 15, 1460–1465 (1997)
12. Zhao, Y.P., Ye, D.X., Wang, G.C., Lu, T.M.: Novel Nano-Column and Nano-Flower Arrays by Glancing Angle Deposition. Nano Letters 2, 351–354 (2002)
13. Dick, B., Brett, M.J., Smy, T.: Controlled growth of periodic pillars by glancing angle deposition. Journal of Vacuum Science & Technology B: Microelectronics and Nanometer Structures 21, 23–28 (2003)
14. Roberts, G.: Langmuir - Blodgett films. Plenum, New York (1990)
15. Reculusa, S., Ravaine, S.: Synthesis of colloidal crystals of controllable thickness through the Langmuir-Blodgett technique. Chem. Mater. 15, 598–605 (2003)
16. Schamel, D., Pfeifer, M., Gibbs, J.G., Miksch, B., Mark, A., Fischer, P.: Chiral Colloidal Molecules and Observation of The Propeller Effect. Journal of the American Chemical Society 135, 12353–12359 (2013)
17. Glass, R., Moller, M., Spatz, J.P.: Block copolymer micelle nanolithography. Nanotechnology 14, 1153–1160 (2003)
18. Baranova, N.B., Zeldovich, B.Y.: Separation of Mirror Isomeric Molecules by Radio-Frequency Electric-Field of Rotating Polarization. Chem. Phys. Lett. 57, 435–437 (1978)

MicroStressBots: Species Differentiation in Surface Micromachined Microrobots

Christopher G. Levey[1], Igor Paprotny[2], and Bruce R. Donald[3,4,5]

[1] Thayer School of Engineering at Dartmouth College, Hanover, NH, USA
[2] Dept. of Electrical and Computer Engineering, University of Illinois,
Chicago, IL, USA
[3] Dept. of Computer Science, Duke University, Durham, NC, USA
[4] Dept. of Biochemistry, Duke University Medical Center, Durham, NC, USA
[5] Duke Inst. for Brain Sciences, Duke University Medical Center, Durham, NC, USA

Abstract. In this paper we review our ongoing research on untethered stress-engineered microrobots (MicroStressBots), focusing on the challenges and opportunities of operating mobile robots on the micrometer size scale. The MicroStressBots are fabricated with planar dimensions of approximately 260 μm $\times$ 60 μm and a total mass less than 50 ng from 1.5-3.5 μm thick polycrystalline silicon using a surface micromachining processes. A single global power delivery and control signal is broadcast to all our robots, but decoded differently by each species using onboard electromechanical memory and logic. We review our design objectives in creating robots on the microscale, and describe the constraints imposed by fabrication, assembly, and operation of such small robotic systems. Our robots have been used to motivate and demonstrate multiple robot control algorithms constrained by a single global signal with a limited number of distinct voltages.

1 Introduction

Microscale mobile devices have many potential applications, including assembly, medicine, and surveillance. The ability to operate multiple microrobots is particularly useful, but challenging to implement using a globally broadcasted control signal. In this paper we review our ongoing research effort on untethered stress-engineered microrobots (MicroStressBots) [1–5], and discuss the challenges and opportunities of operating mobile microrobots at the micrometer size scale. Specifically, we show the application of our robots to controllable microassembly tasks.

Our work is motivated by a goal to develop self-reconfigurable robotic systems at the microscale. This research objective required the development of untethered micro-scale robots with a means of (1) planar locomotion and steering, (2) wireless reception of control and power signals, and (3) on-board control signal decoding (requiring minimal memory), all operating in an environment where multiple robots could interact. To obtain this functionality using a robust mass

I. Paprotny and S. Bergbreiter (Eds.): Small-Scale Robotics 2013, LNAI 8336, pp. 66–80, 2014.

manufacturing process on the micro-scale, design simplicity is key; here the limitations of a simple elegant hardware design are compensated for by more complex control algorithms [4, 5].

2   MicroStressBots

The MicroStressBots consist of a single monolithic plate of polycrystalline silicon with a thin chromium film used to control its out-of-plane shape through stress engineering. Precise design of each microrobot chassis and stress engineering layer ensures the ability to control multiple robots on a single substrate. Fig. 1 shows micrographs of two types MicroStressBots, a single arm design (left) and a dual arm design (right). In both cases, locomotion is accomplished using an untethered scratch drive actuator (USDA) [1], while turning occurs through a snap-down of one of the steering arms.

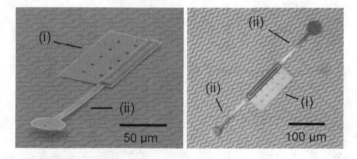

Fig. 1. Micrographs of two MicroStressBots: single arm design (left) and dual-arm design (right). In both cases the untethered scratch-drive actuator (i) provides forward motion, while the steering arm actuator (ii) determines whether the robot moves forward or turns.

All the robots operate on a single power delivery substrate (also called their operating environment). Because of this, a single power and control signal is broadcast over the entire operating environment. Independent control is achieved by differentiating the design of the steering-arm actuators, and thus the behavior of the robots during the application of the global control signal. Fig. 2 shows several MicroStressBots operating on a single substrate.

2.1   Locomotion

Scratch drive (SD) actuation is a well-established MEMS locomotion mechanism (see left panel, Fig. 3). Traditionally, a voltage is applied to the actuator through direct contact using a power rail or a tether wire. Our goal was to implement interacting robots without the constraints of such tethers or tracks, so we devised the capacitive coupling scheme shown in Fig. 3, right panel [1]. Broadcast electrodes are interdigitated uniformly under the entire operating environment, and

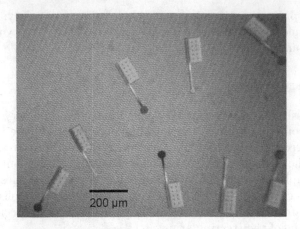

Fig. 2. Multiple MicroStressBots operating on a single power-delivery substrate

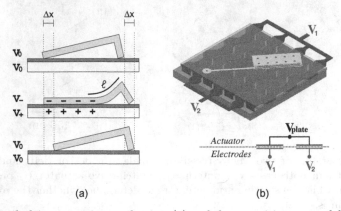

Fig. 3. Scratch drive actuation mechanism (a) and the capacitive power delivery mechanism for untethered MicroStressBots power delivery and control (b)

are powered by a single voltage wire (V_2) and ground (V_1). The electrodes are covered by a thin insulating layer (primarily zirconia). This prevents any direct electrical contact between the electrodes and the robots, and the high dielectric constant (≈ 20) of zirconia enhances the surface charge resulting from applied voltages. This prevents any direct electrical contact between the electrodes and the robots.

A robot positioned on top of this dielectric layer covering several of the interdigitated electrodes will experience a downward force each time a voltage is applied across the electrodes. The scratch drive mechanism illustrated in left panel of Fig. 3 then converts this downward force to lateral motion [6]. For a given applied voltage, the high dielectric constant (20) of the zirconia layer enhanced the surface charges and results in a larger vertical force and hence a stronger SDA locomotion.

2.2   Steering

MicroStressBots use an electrostatic snap-down mechanism for steering as well as locomotion; both actuators are fabricated out of the same layer of doped polysilicon. The turning mechanism is shown in Fig. 4. During pull-down, a portion *s* of the steering arm comes into flat contact with the substrate (Fig. 4.a). When the USDA is subsequently actuated, *s* acts as a temporary anchor, restricting the motion of the tip of the steering arm. The robot follows a curved trajectory, flexing the steering arm until the restoring force of the arm equals the force applied by the USDA (Fig. 4.b). When the arm is released during periodic polarity reversal of the waveform, the flexure in the arm is relieved, resulting in a net change in the heading of the microrobot (Fig. 4.c). The amount of the steering arm flexure is highly dependent on the geometry of the steering arm actuator, making the corresponding turning rate design-specific.

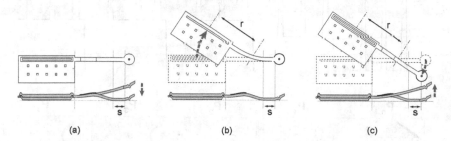

(a)   (b)   (c)

Fig. 4. MicroStressBot turning mechanism

To control the actuation of the steering-arm independently of the scratch drive, we utilize electromechanical hysteresis by designing the arms to respond to different voltage levels. The scratch drive stepping voltages are nested between the steering arm snap-down voltage, V_d, and snap-up voltage, V_u, such that the scratch drive can provide locomotion with the steering arm either up or down. In [2] we show that this nesting is difficult to achieve using ordinary photolithographic patterning, which defines only the in-plane (x-y) shape of the steering arm. However, by widening the design space to include out-of-plane (z axis) geometries, it is possible to incorporate nesting. We integrate such 3D designs into the nominally 2D process of surface micromachining by inducing out-of-plane curvature in the steering arm through stress-engineering: a stressor film is deposited on the arm and patterned in post-processing [7] (i.e. after wafer dicing).

2.3   Species Differentiation

While all our robots receive the same power and control signals, we can vary the design of the steering-arms of the individual MicroStressBots such that different robots respond differently to the global applied control. We call this concept for Global Control, Selective Response (GCSR) [3], and MicroStressBots that exhibit different behavior are said to be of a different microrobot species.

For example, the different steering arm designs (primary varying the length of the arm, the length of the stress-engineering layer, and the pad size) result in distinct threshold voltages for changes in arm state, that is, the snap-down (V_d) and snap-up (V_u) voltages. Each species has a unique V_d or V_u; control signals with different combinations of these will result in distinct motion of the individual devices. Fig. 5 shows five unique waveforms, called control primitives, which are used to differentiate the motion of four distinct microrobot species using differences in the snap-down and release voltages of their steering arms.

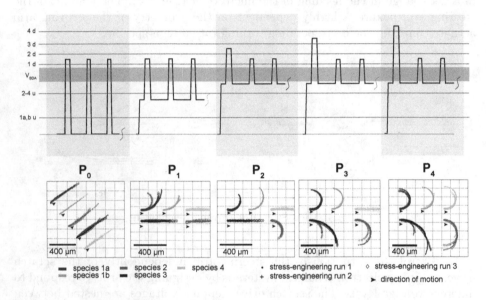

Fig. 5. Differentiation of MicroStressBot species using snap-down and release voltages of the steering arms: Five unique waveform primitives, P_0-P_4 (top) that differentiate the motion of four different species of MicroStressBots (R_1-R_4), distinguished by their snap-down (1d-4d) and up (1u-4u) voltages. Under each control primitive are the experimental trajectories of five robot designs responding to that primitive. Clockwise from top-left of each panel except P_0, they are: R_{1a} (blue), R_{1b} (green), R_4(yellow), R_2(red), R_3(black). Robots R_{1a} and R_{1b} are different designs but belong to the same MicroStressBot species. USDA actuation (flex/release) voltage range (V_{SDA}) is bracketed by arm snap voltages so that MicroStressBots can move with arms up or down. (Based on data from [3]).

It is also possible to differentiate microrobot species using differences in their turning radius rather than snap voltages [4]. Fig. 6 shows the designs (top) and trajectories (bottom) of two MicroStressBot species differentiated by their turning rates. The differentiation stems from design-induced differences in the steering-arm design.

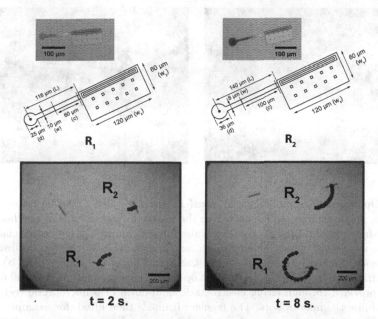

Fig. 6. Differentiation of MicroStressBot species using different turning rates: The design of two different microrobot species R₁ (top-left) and R₂ (top-right), and the trajectories of the robots showing their clearly different turning rates. (bottom) (Based on data from [4]).

2.4 Transfer Frames: Batch Transfer Mechanism for Initial Placement of Robots

In [1] the operating environment and devices were fabricated on the same die along with a self-assembly mechanism, however this co-fabrication imposed severe constraints on the materials and design. In [2] a vacuum microprobe was used to to pick and place single robots on the operating environment, however such sequential manipulation has its limitations with respect to transferring many microrobots onto the operating environment. In [3] we devised transfer frames to enable separate fabrication without imposing a pick-and-place operation, one robot at a time. The operation of the transfer frame is shown in Fig. 7. In this design, twelve robots can be transferred at a time. The robots are fabricated attached to the frame in a set configuration, and maintain their relative placement during the transfer operation. After transfer, the robots can be either immobilized by applying a potential to the electrodes or locked down mechanically using pressure applied by a microprobe. Lifting the frame away from the substrate causes the weak mechanical links between the robots and the transfer frame to be severed, releasing them from the frame.

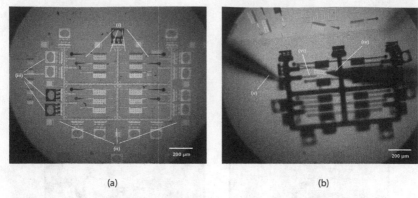

(a)   (b)

Fig. 7. Optical micrograph of a batch transfer structure called a transfer frame. **(a)** Microrobots are manufactured connected to the frame through sacrificial notched beams (i). The frame is anchored to the substrate through another set of sacrificial beams (ii). This second set of beams are broken, and the frame is lifted of the substrate by microprobes inserted into hinged ears at the perimeter of the frame (iii). **(b)** Once a transfer frame is placed on the destination operating environment, mechanical pressure, such as provided through either electrostatic hold-down or a microprobe (iv), is used to immobilize the microrobots. The transfer frame is then lifted, for example, using a second microprobe (v), severing the sacrificial beams (vi) connecting the robots to the frame. (From [3]).

3   Independent Control

Differentiation of microrobot species allows multiple MicroStressBots to be independently controllable within a planar operating environment. Because future microrobtic application will likely rely on simultaneous operation and cooperation of many microrobots, independent control is likely to be an important capability of future multi-microrobotic systems. In [3, 5] independent control has been used to implement planar microassembly, enabling MicroStressBots to independently move into configurations that allows them to dock to form larger structures. The asymmetric friction of the USDAs together with compliance allows the structures to align to a global minimum energy shape through a form of pairwise self-assembly. Fig. 8 shows five structures assembled using species differentiated by different steering arm snap voltages [3]. As described in Sec. 2.3, it is also possible to achieve independent control through other forms of behavioral differentiation, such as variable turning rate [4].

3.1   Global Control Selective Response (GCSR)

The concept of Global Control Selective Response (GCSR) was coined shortly after it became clear that it is difficult to design USDAs to be selectively addressable using the power delivery waveforms [2]. Instead, systems composed of several MicroStressBots were designed to *always* move, but to move *differently*

Fig. 8. Five structures assembled using species differentiated by distinct snap-down and release voltages of the steering arms [3]

during portions of the control voltage waveform. Differentiation through physical design of microrobot chassis is called Global Control Selective Response (GCSR) [8]. Another similar concept called Ensemble Control (EC) was also proposed to be applicable to independent control of underactuated multi-microrobotic systems [9].

In the absence of control error, trajectory planning of independent motion of several microrobots using a single GCSR can be viewed as multiple path planning problem where the motion of the robots is coupled and trajectories of the individual robots are designed (planned) such that all robots enter the desired target configurations at some common point in time. A simple illustration of this concept is shown in Fig. 9 from [5], which shows nominal trajectories of two MicroStressBots that are maneuvered to dock together. The global control signal consists of three control primitives, but only P_2 and P_3 are shown in Fig. 9. Note that the robots motion is differentiated *only* during the application of primitive P_2.

Any physical microrobotic system will experience control error, which will perturb its trajectory, and a control strategy scheme must be devised to minimize such control error. A closed loop re-planning controller can be used, and has been implemented in [3, 5]. Fig. 10 shows the trajectories of four MicroStressBots as they progressively assemble a planar shape. Similar control schemes have been later presented [10].

Finally, compliance [11] between two or more docking microrobots can be used to further reduce the resulting control error. MicroStressBots are specifically well suited for compliant interaction because the USDAs can rotate if the motion on one of its sides is slightly obstructed. This mechanism is used used during turning, however also allows the robots to align during docking. Such self-alignment can be used to remove any residual control error. Fig. 11 shows mutual alignment of two MicroStressBots during docking. The self-alignment reduces final missalignment of the robots at the end of the assembly operation.

Ultimately, independent multi-microrobot control is an important direction of future research. The scalability of the control scheme is of particular importance in order to enable control of future multi-robotic systems composed of large numbers of microrobots. For example, the *SeSAT* control scheme presented in [5] proposes a design methodology that can control n MicroStressBots with sub-linear ($O(\sqrt{n})$) number of control voltage levels.

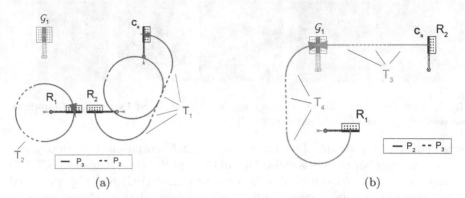

(a)   (b)

Fig. 9. Example trajectory planning for two MicroStressBots maneuvered towards a common configuration. (a) Robot R_2 is maneuvered to an intermediate configuration C_a while robot R_1 follows a circular trajectory (orbit). (b) Robot R_1 is maneuvered to dock with robot R_2, while R_2 moves in straight line. Assuming both robots move at the same speed, the length of trajectory $T_1 + T_3$ is equal to the length of the trajectory $T_2 + T_4$. *Based on data from [5].*

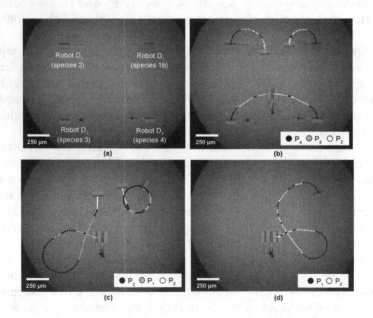

Fig. 10. Progressive assembly of a planar shape through independent control of four MicroStressBots

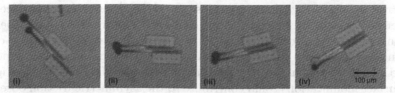

Fig. 11. Optical micrographs taken at the time of initial contact (i) and successively though the alignment process (ii)(iv) showing the self-alignment of two robots

4 MicroStressBot Tribology

USDAs require non-symmetric friction in order to ensure continuous forward motion. Although they move reliably in one direction (as can clearly be seen from Fig. 5), this locomotion mechanism is not yet fully understood. We have observed two surface effects that may contribute to how the MicroStressBots operate on the power-delivery substrate.

4.1 Surface Modification

The USDAs modify the surface on which they move. The left panel of Fig. 12 shows atomic force micrographs of the SiO_2-coated ZrO_2 insulated surface of the power delivery substrate after traversal by a MicroStressBot. Debris pushed forward by the bushing is clearly visible. Scratches in the surface, caused either by asperities or lodged debris, can also be seen.

(a) (b)

Fig. 12. AFM images showing substrate modifications by scratch drive actuation: (a) Visible is the line of debris pushed by the bushing and scratches in the surface caused by either asperities in the bushing or lodged debris. (b) Apparent asperity modification for an area that has be repetitively traversed in one direction (indicated by red arrow). The grey scale corresponds to an elevation of 100 nm (black to white). The x-, and y-scale is in μm.

The right panel of Fig. 12 shows a close-up of the asperities on a surface that has been successively traversed by a scratch drive actuator (SDA). The asperities are elongated in the direction of travel, with a sharper edge on the side corresponding to the forward direction. These apparent changes in surface topology are particulary striking in the 3D visualisation of the AFM data shown in Fig. 13. Such surface modifications could contribute to the asymmetric friction observed. The statistical nature of this process is consistent with recent observations by McGray et. al. [12] showing a distribution of USDA single step sizes.

Fig. 13. A 3D visualisation of the AFM data showing substrate modifications by repeated scratch drive actuation. The division in the x-, and y-scale is 0.5 μm, and the z-scale is exaggerated 10×. The red arrow indicates the direction of travel.

4.2   Surface Charge Injection

The high electric field generated by the power delivery and control signal across the dielectric layer may also cause changes in the environment traversed by a MicroStressBot. Figs. 14 and 15 show scanning electron micrographs of tethered SDAs after actuation over a silicon nitride coated surface. The SEM images reveal a shadow imprint in the substrate, which is likely caused by embedded charges. These silhouettes were only visible in the SEM and not in corresponding optical images. Charging of the substrate was also confirmed by the need for periodic polarity reversal of the power delivery waveform, although the effect was never completely negated. Trapped charges are important because they are likely responsible for variations threshold snap voltages which we have used to set the minimum difference between any two control voltages; this limits the number of distinct control voltages. Even short lived trapped charges could cause hysteretic forces during the course of the SDA actuation, thus affecting the motion of the MicroStressBots.

Fig. 14. Scanning Electron Micrograph (SEM) images of tethered SDA after repetitive actuation over a silicon nitride surface. The shadowy image of the SDA is visible in the underlying substrate does not show up in optical microscopy, and is believed to be caused by charges embedded in the surface). Debris pushed by the bushing is also visible in both images.

5   Scaling Laws at the Microscale

Our MicroStressBots operate in a strictly sub-millimeter size domain. In this section we discuss how scaling laws have influenced their design and fabrication. Although scaling to small size for our thin film devices is not isomorphic, we simplify the dimensional analysis in this discussion by using a single characteristic length scale, parameterized by l [13, 14]. Forces scaling with a smaller power of l become more dominant on the microscale.

A microrobot will generally have a much larger surface (l^2) to volume (l^3) ratio than a large (standard-sized) robot. Effects which scale with mass or volume (l^3), such as inertia and gravitational forces, play a much less important role on the microscale; for example, gravitational forces can be much weaker than adhesion forces. For flat clean surfaces, adhesion forces can scale as surface area (l^2) or perimeter (l) [15, 16]. Physico-chemical adhesion between surfaces typically scales as the area in contact (l^2). In the presence of a condensable fluid, meniscus capillary forces can dominate and scaling then goes as the circumference of the contact area (l), though for a rough surface with multiple asperities in contact, bridges form, so the perimeter of the contact area increases [15] and can become fractal with scaling between (l) and (l^2) [17]. Meniscus capillary forces can work against robot locomotion; we minimize the potentially deleterious effects of such strong forces by operating our MicroStressBots in a dry environment. The fact that a purged dry environment improves their performance is in fact an indication that such forces can indeed be important. Adhesion can also result from built-up electrostatic charges; for constant charge density (or equivalently constant electric field between surfaces), such adhesion also scales as area (l^2), as shown below.

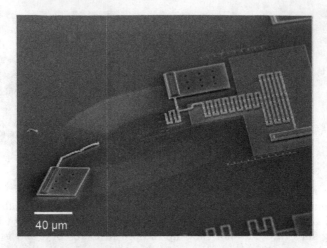

40 μm

Fig. 15. Scanning Electron Micrograph (SEM) images of tethered SDAs after actuation over a silicon nitride surface. The shadow traces left by the SDAs are clearly visible, and are not visible in optical micrographs. After the tether connected to the left SDA broke, the right SDA and the spring were pulled back towards the starting configuration.

Microscale frictional forces typically scale with area (l^2). On millimeter and larger scales, friction depends on load, material, and sometimes velocity, but not on contact size. However on the microscale, adhesion provides a built-in load force which can dominate over the weight or other applied load. For this reason the common assumption for surface MEMS is that friction scales as contact area (l^2) [16]. Air friction also depends on (l^2) [14].

Electrostatic forces for constant electric field also typically scale with area (l^2). For example, the force between plates of a parallel plate capacitor of capacitance C, plate area A, gap x, and fixed voltage V, is given by:

$$F = -\frac{d}{dx}\frac{1}{2}CV^2 \propto A\left(\frac{V}{x}\right)^2 = AE^2. \tag{1}$$

Initially this force is independent of l because as the gap shrinks, the electric field (E) increases, while the area decreases. However in practice, the increasing field is limited by the dielectric breakdown strength of the insulator used (zirconia in our case), and the electrostatic forces then scale with area (l^2). We choose zirconia as our dielectric due to its particular combination of a high dielectric breakdown strength and a large dielectric constant, enabling maximum force with lower applied voltages.

Magnetic forces resulting from constant current density electromagnets scale as l^4 [14] and are thus relatively weaker on the microscale. For example, the magnetic force between two parallel wires of length l, separated by d and carrying currents i_1 and i_2 is proportional to $i_1 i_2 (l/d)$; at constant current density, each current i_1 and i_2 scales as the wire cross-sectional area, so the force scales with l^4. Such magnetic forces are thus relatively weak for small devices. If two permanent magnets are used, scaling is better: for fixed flux density the force between two magnetized surfaces is proportional to area (l^2). While such a system with only permanent magnets could be used for one-time assembly, it does not allow for dynamic control. Microrobots made of hard magnetic materials can successfully interact with external field gradients [18, 19]; the force then scales with magnet volume (l^3), while the relatively large external electromagnets are assumed to remain fixed in size.

A locomotion mechanism based on local electrostatic forces (l^2) is thus attractive for microrobots, both because such forces scale better than magnetics and because they scale in the same way as the dominant dissipation (friction), allowing a range of different sized robots to operate similarly within this regime. Inertial (l^3) and rotational inertia (l^5) effects are less dominant at this scale, simplifying the robot kinematics.

Scratch drives provide perhaps the simplest electrostatic MEMS locomotion mechanism, but traditionally have required a tether or track to provide power. Our thin film capacitive power coupling scheme [1] enables their untethered use. Such harvesting of power from the operating environment is advantageous over schemes involving on-board energy storage because for fixed energy density materials, the storage capacity scales as volume (l^3), making this less attractive at microscale.

An additional constraint for microscale devices is that assembly techniques are limited; our designs are fabricated as a complete unit to avoid the need for assembly. We provide locomotion, steering, power reception, and command decoding all through a simple monolithic structure. Connected transfer frames, such as those described above in Sec. 2.4 also enable the post-fabrication placement of multiple robots as a parallel process.

6 Conclusion

We have fabricated several distinct microrobot species using surface micromachining. Power and control signals are broadcast to the robots from a uniform global environment through capacitive coupling. MicroStressBot species are differentiated by their design, respond uniquely to the same global control signal, and are able to achieve assembly. In this chapter, we have summarised our work on MicroStressBots and introduced some of the many challenges pertained to the simultaneous control of multiple stress-engineered robots at the microscale. Fabrication of mobile microrobots is challenging, however the lack of mass assembly tools at this scale also provides a niche which may eventually be filled using such microrobots. On a micro-factory floor, an army of simple robots could be used to assemble more sophisticated devices out of micromachined parts.

Acknowledgement. This work has been supported in part by the NIH (grant numbers GM-65982, GM-78031, and NS-79929 to B.R.D.), the Office for Domestic Preparedness, Department of Homeland Security, USA (grant number 2000-DT-CX-K001 to B.R.D), the Center for Information Technology Science in the Interest of Society (CITRIS), and the Thayer School of Engineering at Dartmouth.

References

1. Donald, B.R., Levey, C.G., McGray, C., Rus, D., Sinclair, M.: Power delivery and locomotion of untethered micro-actuators. Journal of Microelectromechanical Systems 10(6), 947–959 (2003)
2. Donald, B.R., Levey, C.G., McGray, C., Paprotny, I., Rus, D.: An untethered, electrostatic, globally-controllable MEMS micro-robot. Journal of Microelectromechanical Systems 15(1), 1–15 (2006)
3. Donald, B.R., Levey, C.G., Paprotny, I.: Planar microassembly by parallel actuation of MEMS microrobots. Journal of Microelectromechanical Systems 17(4), 789–808 (2008)
4. Paprotny, I., Levey, C., Wright, P., Donald, B.: Turning-rate selective control: A new method for independent control of stress-engineered MEMS microrobots. In: Robotics: Science and Systems VIII (2012)
5. Donald, B.R., Levey, C., Paprotny, I., Rus, D.: Planning and control for microassembly using stress-engineered. International Journal of Robotics Research 32(2), 218–246 (2013)
6. Akiyama, T., Shono, K.: Controlled stepwise motion in polysilicon microstructures. Journal of Microelectromechanical Systems 2(3), 106–110 (1993)
7. Tsai, C.L., Henning, A.K.: Out-of-plane microstructures using stress engineering of thin films. In: Proceedings of the Microlithography and Metrology in Micromachining, vol. 2639, pp. 124–132 (1995)
8. Donald, B.R.: Building very small mobile micro robots. Inaugural Lecture, Nanotechnology Public Lecture Series. MIT (Research Laboratory for Electronics, Electrical Engineering and Computer Science, and Microsystems Technology, Laboratories), Cambridge (2007), http://mitworld.mit.edu/video/463/
9. Becker, A.T.: Ensemble Control of Robotic Systems. PhD thesis, University of Illinois at Urbana-Champaign (2012)
10. Diller, E., Floyd, S., Pawashe, C., Sitti, M.: Control of multiple heterogeneous magnetic micro-robots in two dimensions. IEEE Transactions on Robotics 28(1), 172–182 (2012)
11. Mason, M.T.: Mechanics of Robotic Manipulation. MIT Press (2001)
12. McGray, C.D., Stavis, S.M., Giltinan, J., Eastman, E., Firebaugh, S., Piepmeier, J., Geist, J., Gaitan, M.: Mems kinematics by super-resolution fluorescence microscopy. Journal of Microelectromechanical Systems 22(1), 115–123 (2013)
13. Trimmer, W.S.N.: Microrobots and micromechanical systems. Sensors and Actuators 19(3), 267–287 (1998)
14. Gosh, A.: Scaling Laws. In: Chakraborty, S. (ed.) Mechanics Over Micro and Nano Scales, ch. 2. Springer, New York (2011)
15. Bhushan, B., Dandavate, C.: Thin-film friction and adhesion studies using atomic force microscopy. Journal of Applied Physics 87(3), 1201–1210 (2000)
16. Williams, J.A., Lee, H.: Tribology and MEMS. Journal of Physics D: Applied Physics 39, R201–R214 (2006)
17. Bora, C.K., Flater, E.E., Street, M.D., Redmond, J.M., Starr, M.J., Carpick, R.W., Plesha, M.E.: Multiscale roughness and modeling of MEMS interfaces. Tribology Letters 19(1), 37–48 (2005)
18. Yesin, K.B., Vollmers, K., Nelson, B.J.: Modeling and control of untethered boimicrorobots in fluid environment using electromagnetic fields. The International Journal of Robotics Research 25(5-6), 527–536 (2006)
19. Pawashe, C., Floyd, S., Sitti, M.: Modeling and experimental characterization of an untethered magnetic micro-robot. International Journal of Robotics Research 28(9), 1077–1094 (2009)

Towards Functional Mobile Magnetic Microrobots

Wuming Jing and David J. Cappelleri

School of Mechanical Engineering
Purdue University, West Lafayette, IN 47907, USA
{jing6,dcappell}@purdue.edu

Abstract. This chapter covers some fundamental work towards realizing functional mobile magnetic microrobots. First, the theoretical fundamentals of electromagnetism are presented. Second, an electromagnetic testbed design for controlling mobile magnetic microrobots is described. It is utilized to perform benchmarking tests on a simple I-bar shaped magnetic microrobot design. After benchmarking, the critical aspects for micro scale robots and two specific microrobot designs are developed addressing the application needs of biomedical and micro manufacturing tasks. They exhibit tumbling and crawling locomotion mechanisms, respectively. Finally, a magnet microrobot body and vision-based force sensor end-effector combination illustrates an approach for combining different technologies together to create the truly functional mobile magnetic microrobots of the future.

Keywords: magnetic microrobotics, micro-force sensing.

1 Introduction

1.1 Motivation

The features and traits of tiny robots hold great promise in biomedical and manufacturing applications. In biology, manipulation at the cellular level is always a critical and challenging task, especially when manipulating live cells during in vivo tasks. Also for the medical applications, there is a desire for tools on the small scale for minimal and non-invasive surgery. Micro scale robots are also needed for advanced micro manufacturing, especially in bottom-up additive manufacturing scenarios. Micro parts typically require operations like handling, sorting, positioning and assembly. Untethered micro agents have the potential to demonstrate advantages in executing these tasks in enclosed incapacious workspaces. Driven by these practical needs, efforts have been conducted towards functional magnetic micro scale robots that are more than pure permanent magnetic bodies and can exhibit advanced functions beyond simple locomotion. The research community is still far away from realizing a micro scale robot that can perform a complex microsurgical or manufacturing tasks with on-board sensing, actuation, and intelligence. However, this allows for many opportunities for new

I. Paprotny and S. Bergbreiter (Eds.): Small-Scale Robotics 2013, LNAI 8336, pp. 81–100, 2014.

ideas and approaches in order to make this a reality in the future. This chapter presents the intial efforts towards creating truly functional magnetic mobile microrobots for biomedical and manufacturing applications.

1.2   Related Work

Since actuation is still a significant challenge for micro scale robots, a major portion of recent research efforts are addressed on the power delivery and working mechanism for microrobots in order to derive mobility and controllability. Autonomous robots on the macro scale are powered by engines, motors, etc., which consume fuel or electricity. At present, these on-board power methods can not be shrunk down to the micro scale, i.e. smaller than 1 mm. Since there are no on-board power sources on that small scale, the representative power solutions of recent work on wireless microrobotics are mainly based on field effects such as electrostatic, thermal and magnetic principles.

The first representative microrobot prototype was presented by Donald et al [1] with the largest dimension of 250 μm. This design applied a scratch-drive actuator (SDA) working mechanism through electrostatic forces. The untethered, electrostatic microrobot consisted of a SDA and a curved, cantilevered steering arm mounted on the actuator. The SDA part was able to move the robot body forward while the steering arm could be clamped down as an anchor to steer the robot's orientation.

Thermal energy is also able to deliver energy over distances wirelessly. A focused laser beam is one of the energetic options for optical driven thermal propulsion [2]. With a focused laser beam, one micro device consisting of a three-legged, thin-metal-film bimorphs structure has been developed by Sul et al [3] with the overall dimension as small as 30 μm. Due to the different thermal expansion coefficient of aluminum and chromium, the leg exposed to the laser beam deflects differently than its original shape after release. Therefore, the device can perform locomotion like an inchworm.

In additional to electrostatic and thermal actuation methods, the magnetic principle also provides a solution for wireless power delivery to a microrobotic agent. Various magnetic microrobot designs have been investigated. On the submillimeter micron scale, the direct method of pulling or propelling a magnetic microrobotic agent uses magnetic field gradients directly. One significant work has been done by Yesin et al [4]. The design was an assembly of two electroplated nickel plates named "OctMag". The assembled soft magnetic agent, of largest dimension of 950 μm, was able to swim in a fluidic environment driven by coaxial Maxwell and Helmholtz coil pairs.

For current microrobotic systems based on the direct propulsion with magnetic field gradients, one considerable limitation is that it requires relatively strong magnetic field flux intensity. This is due to the trend that the magnitude of magnetic force decreases fast comparing with inertia forces when dimensions scale down. The magnetic force also decreases drastically with increasing working distance and agent volume decreases. For the tiny robotic system, it's difficult to set the working distance close and keep the agent volume small at the same time.

These difficulties place a limit on the minimum size of the effective magnetic volume of the microrobotic device. Therefore, other than the direct propulsion by magnetic field gradients, more deliberate working mechanisms are explored for feasible magnetic microrobotic systems.

One of the representative working mechanisms for magnetic microrobotics is based on the oscillation of magnetic bodies. The mobile microrobot presented in [5] is a spring-mass system powered by an external oscillating magnetic field. The robot structure of largest dimension of 300 μm consists of a conductive base frame carrying two smaller asymmetric soft-magnetic masses. The base frame is made of gold which is non-magnetic while the magnetic masses are made of nickel that serve both as a magnetic "attractor" and a mechanical stopper. This resonant oscillation propels the agent to move in a desired direction in the plane due to the asymmetry of the "attractors".

Another representative series of work on magnetic microrobotics had been done in Sitti's group at CMU [6]. The agent is laser cut from a metal sheet with the largest dimension of 250 μm. The metal sheet is made of permanent magnetic material, NdFeB, that has built-in polarization and retains constant magnetization. This feature makes it possible to predict the agent response when exposed to time-varying magnetic fields. A sawtooth-shaped field signal was generated to induce a cyclic rocking action that resulted in stick-slip motion on various surfaces.

For the dry surface but not limited to it, Hou et al. [7] designed a rolling locomotion method for a magnetic microrobot. The rolling magnetic microrobot was fabricated by dripping adhesive onto an iron wire. A micro ball with a diameter of 440 μm was formed after minutes of drying due to the cohesive force. An external rotating magnetic field was generated by a rotating permanent magnetic block underneath the working substrate. The magnetic force along with normal blocking and friction force enabled successive rotations and locomotion.

1.3 Roadmap

The magnetic principle has been chosen here to actuate the micro scale robots based on two major concerns: (1) It is a convenient way to realize untethered power delivery over distance; (2) It does not need an engineered environment, which is usually not available in the biomedical working environments nor in small enclosed workspaces.

Once the power principle has been chosen, the required magnetic coil test-bed and control module are developed to provide customizable magnetic fields with required intensity level (Section 3). Initial studies begin with an I-bar shaped robot design to derive the power input and resistance force levels on the micro scale. The power input to the agent can be evaluated through calculating the intensity of the magnetic field and the magnetization of the agent volume. The undetermined part is the resistance force and power consumption at the micro scale, such as the friction resistance, electrostatic adhesion, etc. These forces are critical to benchmark the microrobot's behavior prior to the future development (Section 4).

The micro magnetic tumbling microrobot (μTUM) design aims to be capable to fulfill tasks in biomedical applications. For the micro scale robots, the primary features of biological and medical environments are: (1) complex, variable resistance and damping; (2) 3D surfaces or obstacles. Therefore, the microroobt needs adaptable mobility with limited power levels. The philosophy here is to use the lowest power to realize as much possible mobility on variable surfaces and medias, which is the reason that tumbling working mechanism is investigated on the micro scale (Section 5).

Inspired by types of stepwise locomotion mechanisms in nature, a crawling microrobot design applying the magnetostrictive principle is explored to achieve stable incremental motions on dry surfaces. This feature accommodates the advanced manufacturing task such as precise positioning and assembly of micro objects. The manufacturing environment is usually dry and the objective requires stable and precise locomotion. Thereafter, a magnetostrictive asymmetric bimorph (μMAB) microrobot design addresses applications in advanced micro manufacturing scenarios (Section 6).

Since in-situ sensing is necessry for truly functional robots, the micro force sensing function is incorporated into the wireless microrobot design as a microforce sensing mobile microrobot ($\mu FSMM$) (Section 7). Although the micro force sensing function will not complete the whole roadmap to functional magnetic microrobots, it can be used as launching pad to more functional designs, such as ones with on-board micro and nano actuators and other sensors.

2 Theoretical Fundamentals

2.1 Magnetic Phenomenon

Maxwell's equations build up the original foundation of classical electromagnetism. This classical set of four relationships is derived into equations that can accurately predict the electro-magnetic behavior larger than quantum level. Micro scale magnetic robots are larger than quantum dimensions, therefore, the equations are still valid for the magnetic phenomenon on the micro scale robots.

Although Maxwell's equation set itself is a complete description, Ampère's circuital law and the Bio-Savart law provide the basic concept for the magnetic phenomenon. The magnetic field density vector $\mathbf{B}$ inside a magnetic body not only depends on the magnetic field but also on a superposition with the body's magnetization which is related to the material's magnetic property itself. The $\mathbf{B}$'s unit is Tesla (T) while the pure magnetic field intensity vector $\mathbf{H}$ is measured in Ampères per meter (A/m). The relationship between $\mathbf{B}$ and $\mathbf{H}$ is derived through:

$$\mathbf{B} = \mu_0(\mathbf{H} + \mathbf{M}) = \mu_0(\mathbf{H} + \chi\mathbf{H}) = \mu_0(1 + \chi)\mathbf{H} \tag{1}$$

where $\mathbf{M}$ is the magnetic body's magnetization vector and χ indicates this magnetization ability called magnetic susceptibility. The item $1 + \chi$ is termed as μ_r, which is named relative permeability of the material. A core of high permeability μ_r is usually inserted inside a solenoid to strengthen the magnetic field $\mathbf{B}$ by μ_r

times accordingly. Note that the magnetic properties are not only determined by the material composition but also by the physical form like the crystalline phase. Therefore, it might be more complex to interpret the magnetic behavior at the micro scale due to different forming methods when compared to traditional macro-scale manufacturing process.

2.2   Magnetic Force

Any magnetized body within a magnetic field will experience force and torque, which is the actuating basis for the magnetic microrobot designs. The magnetized body always has the tendency to align its internal magnetization according to the streamline of the external field. In any case, the acting force on the magnetic body is in the gradient direction of the magnitude of the applied magnetic field.

If the magnitude of the inner field strength is treated as a constant value for simplification, the acting force and torque can be derived by:

$$\mathbf{F}_m = V_m \, (\mathbf{M} \cdot \nabla) \mathbf{B} \tag{2}$$

$$\tau_m = V_m \, \mathbf{M} \times \mathbf{B} \tag{3}$$

where $\mathbf{F}_m$ and τ_m are the acting force and torque on the magnetic body respectively, and V_m is the volume of magnetic part. Therefore, it is apparent that three factors play primary roles in the exerting forces and torques on a magnetic body, where the volume of magnetic material scales linearly.

3   Experimental Setup

Based on the electromagnetic theory, most of the artificial magnetic fields are produced by electromagnetic coils whose common form is a solenoid, an iron core wrapped with conducting wires. The resulting overall magnetic field intensity vector can be derived through the algebraic sum of multiple solenoids' contribution.

Thereafter, in order to actuate the microrobot designs, a first edition testbed consisting of five independently controlled solenoid coils has been constructed for customizable magnetic field signals (Figure 1(a)). One coil is built with more turns than the others and mounted as the bottom coil to provide the vertical magnetic field. The other four coils are manufactured identically with the same dimensions and number of turns to produce the horizontal magnetic fields. All five coils have cobalt-iron cores inserted with high magnetic permeability which increases the field strength. The work space area is encompassed by the four core end faces in a 1" square. The resultant field has been assessed and calibrated with measurements from a DC Gaussmeter and simulation in COMSOL software.

The second edition compact coil system has also been manufactured consisting of six coils in a more compact setting (Figure 1(b)), where the footprint of total coil system is under 6". The four side coils are still in solenoid form with

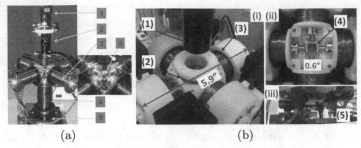

(a)   (b)

Fig. 1. (a) Photograph of the electromagnetic test bed. (1) CCD camera; (2) Micro-scope lens; (3) One of the four side coils set for the horizontal magnetic field; (4) Bottom coil for the vertical magnetic field; (5) X-Y stage; (6) Chamber in workspace. (b) Photograph of the 2nd generation, compact electromagnetic test bed. (i) Overview photograph of the compact test bed: (1) Overhead CCD camera; (2) Top coil of the Helmholtz coil pair; (3) Side coil produces horizontal magnetic field. (ii) Overhead view without the top Helmholtz coil: (4) Workspace. (iii) Bottom view of the coil system: (5) Bottom coil of the Helmholtz coil pair.

cobalt iron cores inserted and they form a 0.6" square workspace. The upper and bottom coil are wrapped with less turns in order to fit the compact dimension. This coil pair in the vertical direction is in a Helmholtz setting without strengthening cores. This is for uniformly distributed vertical magnetic field that has no field gradient across the horizontal plane. The setting also makes it possible for observation with an overhead camera along the vertical axis.

Auxiliary hardware includes imaging, control and power supply systems. The real-time imaging is accomplished with an overhead CCD camera along with a microscope lens of adjustable magnification. The control commanding signals are sent from a PC GUI program written in Labview software. The coils are powered by a two channel variable power supply, where a matched drive circuit is used to amplify the control signal.

4   Magnetic Microrobot Performance Benchmarking

So far, for the untethered submillimeter microrobots, mobility and adaptability are still not resolved to perform functional tasks for real applications, such as in biomedical or advanced manufacturing scenarios. The adaptability and mobility are not separate but also not in a monotone relation either; faster speed does not indicate better adaptability in many cases. While inertia becomes less significant when the dimension decreases to the micro scale, the robot's performance at a particular instant does not only depend on its working mechanism but also the environmental factors. Therefore, the first logical step toward functional magnetic microrobots is to benchmark the various environmental forces that a microrobot will experience. These are summarized as "resistance", although they may not always play negative roles. Since the acting magnetic force can be directly evaluated through the principle, these resistant micro force entities are essential to benchmark the adaptability and mobility of the microrobot.

4.1   Resistance

The micro scale robots are primarily susceptible to surface force on dry surfaces and fluid drag in fluidic environments. For the submillimeter microrobot, the resistance on dry surface can not be simply estimated by the friction force due to weight, since the inertial forces no longer play the dominant role at this small scale. Instead, the resistance here on a dry surface is considered to include adhesion forces.

The primary source of the "adhesion" comes from the electric charge accumulated on the object's surface. The magnitude of the electrostatic force F_e can be derived based on *Coulomb's law*:

$$F_e = \frac{1}{4\pi\epsilon_0} \frac{q_1 q_2}{r^2} \tag{4}$$

where ϵ_0 is the vacuum dielectric constant, $K_e = 1/4\pi\epsilon_0 = 8.987 \times 10^9 \, Nm^2 C^{-2}$ is called *Coulomb constant*, q_1, q_2 are the interacting charges and r is the distance between them. Referring to the data of charge density, the electrostatic force acting on a micro scale agent can be accumulated up to more than 10 μN.

In fluidic environments, the "resistance" on an object is generated by a pressure gradient and friction drag on the object surface. As long as our study object is small compared to the environment, the pressure gradient can be set as zero. Hence the fluid drag F_D is able to be written as:

$$F_D = \int_S \tau dA \tag{5}$$

where S indicates the object surface, A is the total surface area in contact with the fluid, and τ is the shear stress. The fluidic drag force is usually evaluated through empirical drag coefficients, related to the Reynolds number, Re.

4.2   Experimental Evaluation of Resistance

In order to experimentally evaluate the resistance force on micro scale, a design of a soft magnetic material volume is proposed and fabricated, whose geometric contour is outlined by a MEMS processed layer (Figure 2). Thus, it is straightforward to derive the power level that the micron scale magnetic agent owns. Furthermore, it is possible to extract the micro resistant force information through motions with customized parameters in various environments.

The experimental method to derive the adhesive micro force on the dry surface is conducted with a micro force sensor produced by FemtoTools [8]. The force sensor chip is screwed onto a mounting plate attached to a shaft. The shaft is further fixed to a manipulator which can move in the $X - Y - Z$ directions with 65 nm step sizes (Figure 3). I-Bar prototypes with various aspect ratios and

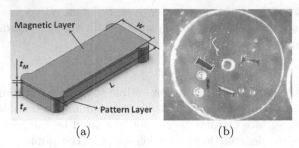

(a)   (b)

Fig. 2. I-Bar shape microrobot design driven by magnetic field gradients. (a) CAD model of I-Bar shape magnetic microrobot. (b) Prototypes of I-Bar shape magnetic microrobots in different geometric aspect ratios.

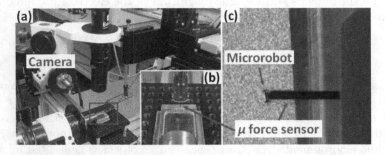

Fig. 3. Experimental setup for resistance on dry surface. (a) Overview of the experimental setup. (b) Local view of the workspace and the force sensor tip. (c) Overhead view of the pushing test under the microscope.

different magnetic layer thickness are pushed by the force sensor (Figure 3(c)). The tests are conducted on both sputtered gold and unpolished silicon surfaces. The root mean square (RMS) values of their surface roughness are in nm and μm magnitude, respectively.

These test results show that the resistance force on dry surface for the small agent ranges from a few μN to several hundreds of μN. This fact indicates that the normal friction becomes trivial on this small scale, which is due to the inertia force such as weight. This trivial amount is governed by the normal blocking force and coefficient of friction. Except the above amount of friction, adhesion is also observed during the tests. This adhesion includes the attraction due to electrical charge. Based on Equation 4 and the test result, the electrostatic force is evaluated to be up to μN levels in magnitude. This is larger than the magnetic force exerted on the micro scale magnetic agent, which is in nN range. This conclusion corroborates the fact that the existing adhesion dominates the micro scale magnetic robot in a field intensity in the mT range.

The test strategy for fluid drag is to record the swim velocities of the I-Bar micro agent in both water and oil environments. The fluid test is conducted in a 1" cube chamber fit into the workspace of the coil system. Since the fluid is contained in a small chamber, the capillary effect and surface tension will lead to a slight curve of the water surface that impacts the translation velocity and drag force on the microrobot body. Additionally, the orientation of the agent also has influence on the velocity. In order to cross out the unrelevant factors, four different current inputs are tested for each type of microrobot agent. The three bigger current inputs and velocity outputs are subtracted by the minimum reference set to cross out the surface effects. Therefore the rules affected by the fluid drag and corresponding parameters can be extracted from the series of the velocity differentials. An example result is shown in Figure 4. The tests and following calculations indicate that the microrobots are in a stable laminar flow domain in both fluids with low and high viscosities. The fluid drag force is determined to be on the order of nN.

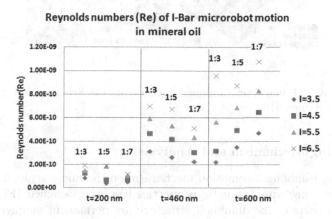

Fig. 4. Reynolds number of motion tests in fluid. The media is mineral oil with kinematic viscosity of $350c.s.t.$ in room temperature. t is the thickness of the magnetic layer on the magnetic microrobot. The fraction indicates the geometric aspect ratio of the I-Bar shape. I indicates the input current to the drive coil.

Evaluation results of the micro resistant forces provide a beneficial reference for other future magnetic microrobot designs on the micro scale, such as the necessary magnetic volume or required field intensity which is valuable for the auxiliary hardware development. On the other hand, this preliminary information is useful to estimate the adaptability and mobility of the microrobot in various complex working environments.

5   The Tumbling Magnetic Microbot (μTUM)

Applications in biology and medicine is one of the primary drivers for mobile microrobot research. Real bio-environments are usually not flat but complex surfaces of tissue, flagella, etc. The evaluation of micro resistant forces indicates that various types of forces on the surface are the major portion of the challenging resistance for the microrobot. Thus, a working mechanism that decreases the contacts with surface would be beneficial. However, totally lifting the robot up from surface and removing all friction will make the agent exhibit fast uncontrollable behaviors. The philosophical trade-off here is to take advantage of the surface friction but reduce the surface contact as much as possible. Therefore, a tumbling motion style could be the solution for an adaptive working mechanism. The tumbling magnetic microrobot, μTUM, (Figure 5) addresses this need to be able to negotiate complex surfaces in biomedical environments [9–11].

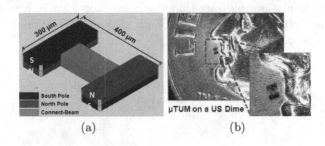

(a)   (b)

Fig. 5. The μTUM magnetic tumbling microrobot. (a) Schematic of μTUM. (b) μTUM on a US dime. [10]

5.1   Working Mechanism and Analysis

To realize the tumbling locomotion mechanism on the micro scale, a composite dumbbell structure with magnetic properties has been designed (Figure 5(a)). The two bell parts of the dumbbell structure are permanent magnets with opposite polar directions. The bell ends are connected by a non-magnetic bridge part.

Suppose the microrobot body lies on the working surface (Figure 6(a)). When the magnetic field in upward ($+z$) direction is turned on (Figure 6(b)), bell A will be pulled down whereas bell B will be repelled up. This pair of forces will generate a pure moment (force couple). If we turn on the horizontal magnetic field pointing right ($+x$) at the same time when turning off the vertical field (Figure 6(c)), the dumbbell device will experience a continuous moment making itself tumble forward (Figure 6(d)). If the device needs to tumble to the left ($-x$), all that needs to be done is commanding the magnetic field signals in the opposite direction. The beauty of this tumbling mechanism is the adaptability to different non-idealized surfaces which are common in biomedical environments. During this locomotion process the surface is not necessary to be ideally flat or horizontal as long as it has contact between the microrobot and the substrate.

Another sliding operating mode is also able to be accomplished by this tumbling microrobot design. When in a flat local area, the tumbling microrobot is able to translate in a simpler standing up-sliding locomotion mode (Figure 7). After the μTUM agent stands up, one can turn on the horizontal field in a pulse of certain length oppositely (Figure 7(c)) compared with the step(c) in tumbling cycle (Figure 6(c)). Therefore, the lower bell part will be pulled resulting in a

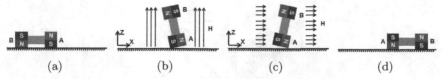

Fig. 6. Tumbling motion mechanism of the magnetic tumbling microrobot. (a) Initial position; (b) Apply vertical magnetic field to make the agent stand up; (c) Apply horizontal magnetic field to fulfill the tumbling locomotion cycle; (d) Final position of one tumbling cycle.

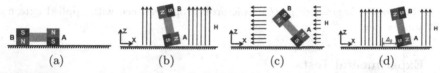

Fig. 7. Sliding motion mechanism of the μTUM microrobot. (a) Initial position; (b) Apply vertical magnetic field to make the agent stand up; (c) Apply horizontal magnetic field to pull the lower bell part for sliding; (d) Turn off the horizontal field while keep the vertical field on, the agent will stand in a translated position.

sliding motion (Figure. 7(d)). Only repeating the last two steps can accomplish the sliding locomotion cycles. This sliding locomotion is not as adaptable as tumbling is to complex environments. Since it has relative motion between the microrobot body and the surface. However, it is more suitable for certain manipulation tasks after the agent reaches the goal area through tumbling locomotion.

The general case of a force analysis model for the tumbling robot is shown in Figure 8, where the arrows in orange color indicate the directions of the magnetic field on the magnetization of the bell parts. The black arrows show the forces and torques that act on the agent. φ_A and φ_B are the angle contained by the vertical direction of the streamline of magnetic field at position A or B, respectively, whereas θ is the agent's incline angle from the horizontal surface. Based on d'Alembert principle, the agent's equilibrium stance can be evaluated through:

$$\frac{\delta W}{\delta \theta} = V\, M\, \nabla B_B\, cos^2\theta\, L - \frac{1}{2}(G + F_a)\, L\, cos\theta + V\, M\, (B_B - B_A)\, sin\theta = 0 \quad (6)$$

The model analysis indicates that a larger exerting magnetic force is helpful for magnetic actuation, including this tumbling design. While a larger volume is not feasible on micro scale, larger field gradients along the vertical direction is beneficial for the performance. Moreover, with the same magnetic force, this tumbling design can conquer larger adhesion than driving the magnetic agent by field gradients directly. The performance can also be enhanced by a completely uniformly distributed vertical field which means less field gradient across the horizontal plane.

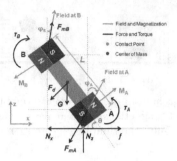

Fig. 8. Free body diagram of μTUM microrobot on a surface with applied external forces and torques

5.2   Experimental Tests

This tumbling magnetic microrobot design has been fabricated through a custom surface MEMS process [10]. The three parts are patterned through photolithography with negative photoresist step by step. The magnetic bell parts are cast by the photoresist mixing with permanent magnetic powder. It is polarized during the soft bake process before exposure. The prototypes have shown the opposite polarization successfully. The tumbling locomotion mechanism has been verified through manually controlled signal (Figure 9).

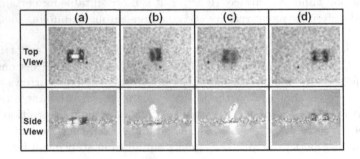

Fig. 9. Steps in one cycle of a μTUM agent performing tumbling motion. The states of top views and side views are captured in different cycles of tests when a side camera is temporarily set up.

The motion adaptability has been tested on various substrates, such as paper, glass, bio-tissue, and rough surfaces like a U.S. penny (Table 1). The tests show that the μTUM microrobot owns about the same mobility on different surfaces, which confirms the motion adaptability of the tumlbing locomotion mechanism at the micro scale.

Table 1. Results of tumbling tests on various substrates

Substrate	Travel distance (mm)	Time (s)	Velocity (bodylength/s)
Paper	1.76	4.3	1.02
Glass	1.64	4.5	0.91
Penny	3.45	9.4	0.92
Tissue	2.22	5.7	0.97

Note: The tests have been run with manual control.

The sliding locomotion and manipulation ability of this magnetic tumbling microrobot has also been verified with the proposed field signal sequence (Figure 10). The manipulation force was derived with pushing tests against a tip of an atomic force microscope tool and determined to be approximately 4 μN.

Fig. 10. Pushing manipulation test of μTUM prototype working in sliding mechanism. A triangle peg is pushed within one sliding locomotion cycle and released. This test is also conducted in an oil bath (viscosity = 40 c.s.t.).

6   The Crawling Magnetic Microbot (μMAB)

Untethered micro-scale end-effectors for advanced manufacturing on the micro scale will be advantageous for additive manufacturing operations in tight enclosed workspaces. The requirements of advanced manufacturing tasks are typically precise positioning and stable motion, needed for assembly and manipulation tasks. Therefore, the logical idea for a microrobot fitting the requirements is to exhibit incremental locomotion on a dry surface. Thus, a crawling **micro**-scale **M**agnetostrictive **A**symmetric thin film **B**imorph (μMAB) microrobot design has been proposed and investigated (Figure 11).

6.1   Working Mechanism and Analysis

This **micro**-scale **M**agnetostrictive **A**symmetric thin film **B**imorph (μMAB) microrobot [12, 13] consists of a magnetic film bonded to a nonmagnetic substrate. Due to the magnetostrictive phenomenon, stress is produced in the film when it's exposed to magnetic field. Bending occurs if one end of the two layer structures

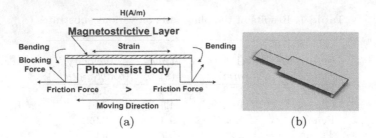

Fig. 11. (a) Actuation principle of the μMAB magnetic microrobot; (b) Isometric view of the μMAB schematic

is clamped. Further, if the deflected end is in contact with some ground or face, a blocking force is produced which is able to provide mechanical work through the friction force it causes. Legs with different geometry dimensions and different contact lines/areas are able to lead to different blocking forces and then friction forces. Making use of the friction difference along the contact face between the robot and the ground can push or pull the robot mass in an incremental step.

Few theories and software tools are able to simulate and predict a planar magnetostrictive bimorph's behavior. What has been done here is translation of the magnetostrictive problem to a piezoelectric problem, because the later situation has analysis tools available. Essentially the same as the piezoelectric phenomenon and its converse effect, piezomagnetic and magnetostrictive effects are opposite phenomena. The piezomagnetic principle is described as:

$$\begin{cases} \varepsilon = \frac{\sigma}{E_y^H} + d_{33}^\sigma \mathbf{H} \\ \mathbf{B} = d_{33}^{H^*} \sigma + \mu^\sigma \mathbf{H} \end{cases} \tag{7}$$

where ε is strain, E_y^H is Young's modulus at constant magnetic field $\mathbf{H}$, $\mathbf{B}$ is magnetic induction, μ^σ is permeability at constant stress, d_{33}^σ is its the axial strain coefficient while $d_{33}^\sigma = d\varepsilon/d\mathbf{H}$, and $d_{33}^{H^*}$ is its inverse coefficient while $d_{33}^{H^*} = d\mathbf{B}/d\sigma$.

For the piezoelectric phenomenon, the so-called coupled equations illustrating the principle are described as:

$$\begin{cases} S = [s^E]T + [d^t]E \\ D = [d]T + [\epsilon^T]E \end{cases} \tag{8}$$

where S is strain, s is compliance stiffness and T is stress. D is the electric charge density displacement (electric displacement), ϵ is permittivity and E is zero or constant electric field strength, d is the matrix for direct piezoelectric effect and d^t is the matrix for the converse piezoelectric effect. The superscript E indicates a zero or constant electric field; the superscript T indicates a zero, or constant, stress field; and the superscript t stands for transpose of a matrix.

One can compare the items of the piezomagnetic and piezoelectric equation sets correspondingly (Equation set 7 and 8). The units of the parameters are

Table 2. Analogies in Piezomagnetic and Piezoelectric Domains

	Magnetostriction	Piezoelectricity
Physical quantity	Strain(ϵ)	Strain (S)
Unit	1	1
Physical quantity	Elasticity($1/E$)	Elasticity(S_E)
Unit	$(\frac{N}{m^2})^{-1}$	$(\frac{N}{m^2})^{-1}$
Physical quantity	Stress (σ)	Stress (T)
Unit	$\frac{N}{m^2}$	$\frac{N}{m^2}$
Physical quantity	Magnetostriction (d)	Piezoelectricity (d)
Unit	$\frac{V \cdot s}{N} = \frac{m}{A}$	$\frac{C}{N} = \frac{A \cdot s}{N} = \frac{m}{V}$
Physical quantity	Magnetic field (H)	Electric field (E)
Unit	$\frac{A}{m}$	$\frac{N}{C} = \frac{V}{m}$
Physical quantity	Magnetic induction (B)	Electric displacement (D)
Unit	$\frac{N}{A \cdot m}$	$\frac{C}{m^2} = \frac{N}{V \cdot m}$
Physical quantity	Permeability(μ)	Permittivity (ϵ)
Unit	$\frac{H}{m} = \frac{A \cdot s}{m \cdot V}$	$\frac{F}{m} = \frac{V \cdot s}{m \cdot A}$

Table 3. Parameters and results of magnetostrictive cantilever

Material	W	L	t_m	t_s	H (V/m)	δ_{sim}	δ_{cal}	F_b (μN)
Nickel	280	190	1	10	2×10^5	0.036	0.037	6.376
Terfenol-D	280	190	1	10	2×10^5	5.921	5.976	629

Note: W is the width, L is the length, t_m is the thickness of the magnetostrictive layer, t_s is the thickness of non-magnetic substrate layer, H is the exterior magnetic field, δ_{sim} indicates the simulated deflection in COMSOL software, δ_{cal} indicates the calculated deflection, and F_b is the blocking force. All the geometric items are quantified in μm.

summarized as Table 2. We can see that the calculation and effects of magnetostriction can be mapped to the piezoelectric domain. The mathematics of piezoelectricity and magnetostriction are essentially the same if the effect is treated as a one-dimensional property. Based on the design parameters [12, 13], with an operating current of 5 A the deflection of a nickel bimorph layer is simulated as approximately 5 μm and the corresponding blocking force is calculated at about $6\mu N$ (Table 3).

6.2   Experimental Tests

This μMAB design has been fabricated with custom MEMS techniques including photolithography and electroplating [13]. The finger structure and body film are patterned layer by layer. The top magnetostrictive layer is electroplated last. The current magnetic layer is made of nickel since it can be easily deposited through electroplating. However, its magnetostrictive property is much lower than the composite magnetic material named "Terfenol-D", which is ideal for this application but can not be processed with electroplating.

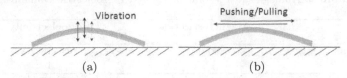

Fig. 12. Two actuation modes of μMAB prototypes. (a) Vibration actuation from oscillating field. (b) Pulling actuation by magnetic field gradients.

For the μMAB made from nickel, the actuation is coupled with both magnetrostrictive and magnetic phenomena (Figure 12). The desired actuation mode for μMAB prototypes is shown schematically in Figure 12(a).

An input of 20 current pulses of 5 A are incrementally applied to the side coil pair in a frequency of approximately 4 kHz. The pulse signal causes the robot to vibrate/deflect and translate across the substrate. At the conclusion of the pulse train, the robot motion ceased. Upon the application of another magnetic field pulse train, robot movement resumed in a similar manner. Snap-shots from one such test illustrating this actuation mode are shown in Figure 13. It is observed from the experimental results that the expected magnetostrictive phenomenon is coupled with the strong magnetic field gradient, which leads the robot body to translate in the lateral direction.

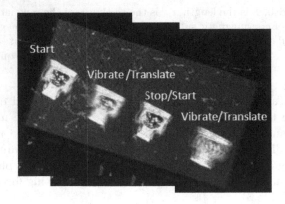

Fig. 13. Snap shots of vibration mode translation of μMAB prototype

Therefore, it has not been determined how the magnetized body and magnetostrictive principle of the body are coupled and interact at the micro-scale. More accurate theoretical modeling of the planar magnetostrictive bimorph, considering the initial magnetization is necessary for precise prediction of the microrobot's behavior based on magnetostrictive principle. However, this design does show great potential for realizing stable incremental motion on dry surfaces for advanced manufacturing applications.

7   The Micro-Force Sensing Mobile Microrobot (μFSMM)

For the micro scale robot, the evironmental conditions are critical for the micro-robot design and performance. Therefore, the effort here is to develop a magnetic microrobot design with a sensing module that is able to provide in-situ micro force feedback wirelessly. This Micro-Force Sensing Mobile Microrobot ($\mu FSMM$) joins a magnetic microrobot body part and a non-magnetic, vision-based force sensing end-effector (Figure 14(a)). The magnetic part is the driven part of the robot that is manipulated with magnetic fields in the workspace. The force sensor part works as a vision-based force sesor that provides micro force feedback, without wire, circuit or any complex electronic device [14]. From knowledge of the planar stiffness of the mechanism and observing the displacement with an overhead camera as it interacts with objects, the forces can be determined.

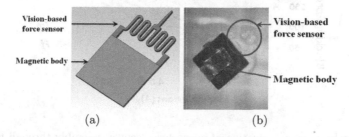

(a)   (b)

Fig. 14. (a) μFSMM design. The base part in pink color indicates the magnetic drive part. The spring structured part in greem color shows the vision based force sensor module. (b) Prototype of μFSMM design.

The design parameters of the micro force sensor module can be found in [14]. The stiffness parameters are derived from finite element simulation and AFM indentation tests of fabricated prototypes. It shows that the force sensor has different stiffness in the X and Y directions.

The force sensor part is made of polydimethylsiloxane (PDMS) for low stiffness. In order to manufacture the micro force sensor part, a micro mold made from negative photoresist is patterned in advance. The released force sensor part is bonded with beryllium copper piece for attachment with a magnetic body piece (Figure 14(b)).

A prototype $\mu FSMM$ has been experimentally tested with the first edition coil system. The input current in the coil is gradually increased from $1A$ to $6A$ to drive the robot into a fixture for initial testing. Increasing the current input results in larger field gradients and also larger deflection of the micro force sensor due to the larger blocking force. The evaluation of the blocking forces through the two methods are compared in Figure 15. In the first method, the blocked force, F_m, is determined by calculating the exerting magnetic force on the robot based on the input current level and magnetic body volume and properties.

In the second method, the blocked force, F_f, is determined from experimentally observing the deflection of force sensor and multiplying this value by the stiffness of the mechanism in the appropriate direction. The results for each method show good agreement, with errors in the tens of nN range (same order as the force sensor resolution). Therefore, we can set up a monotone relationship between the current input and output force on the tip of the micro force sensor robot. Thus, the vision based force sensor on a magnetic microrobot body is able to provide reliable force feedback information to a teleoperator or to a force-guide closed loop control system.

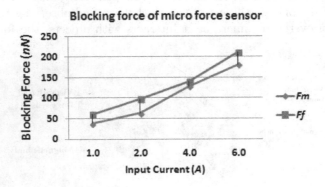

Fig. 15. Blocking force evaluation of micro force sensor assembly through two methods. F_m indicates the experimental result from applying the magnetic drive force. F_f indicates the evaluation result through multiplying the tip deflection by known stiffness.

8   Discussion and Conclusions

Actuation and mobility are still the major challenge for micro scale robots. These magnetic microrobot designs address the difficulties of practical applications. This work provides a beneficial starting point for the truly functional submillimeter magnetic microrobots of the future for biomedical and advanced micro manufacturing applications.

The submillimeter scale magnetic microrobot is able to generate a small amount of acting force ($[nN , \mu N]$) when compared to the environmental resistance that it must overcome. To address this difficulty for the mobility of the magnetic microrobots, the resistance exerted on the micro agent is explored in both dry and fluidic environments. An I-Bar shape magnetic prototype has been designed and fabricated as the end-effector to explore the range of the resistance at micro scale. It is confirmed that inertia force plays a minor role (nN) while stochastic adhesion and electrostatic forces are significant ($[nN , \mu N]$). In the fluidic environment, the micro scale agent moves in the laminar flow in both high and low viscosity. The micro force data provides benchmark information for future microrobot system development in specific environments.

The μTUM microrobot is able to work in a tumbling motion mechanism driven by a predefined sequence of external magnetic fields. By switching the control signal during the cycle, a sliding locomotion mode can also be realized with advantages for executing manipulation tasks. The most appealing advantage of this μTUM microrobot design is the adaptability to complex environments and flexible operation modes, which will provide some beneficial mechanisms for effective tools in real biomedical applications in future, such as drug delivery in hard to reach areas or scraping and pushing tasks at micro scale. This tumbling mechanism at the micro scale can be further explored with more uniformly polarized prototypes. The coil testbed and control system can also be improved for better magnetic field distribution.

The μMAB microrobot design has been verified by modeling and analysis with translating the piezomagnetic principle to the piezoelectric domain. The behavior of prototypes made from nickel indicates that magnetic and magnetostrictive principles are actuating the microrobot with two operating modes. A more complicated and functional microrobot design should be feasible by suitable design and application of structures using micro-scale magnetostrictive mechanisms. The results also indicate that a more realistic theoretical model and calculations are essential to capture the coupling effect of the magnetization and magnetostrictive phenomenon for the micro scale magnetic robot.

The $\mu FSMM$ design provides a concept of a micro scale robot accomplishing an in-situ task wirelessly. It incorporates the micro force sensing function into the mobility of the magnetic microrobot. The sensed force of the current prototype is in nN range.

The work presented here shows that the most critical aspect of magnetic microrobot development is the limited power level of the robots when compared with the environmental resistance that they must overcome. A delicate working mechanism is required based on magnetic principles and it should be tailored to specific application scenarios. The tumbling and crawling working mechanisms have shown merit for functional micro scale magnetic microrobots. The magnet microrobot body and vision-based force sensor end-effector combination illustrates an approach for combining different technologies together to create the truly functional mobile magnetic microrobots of the future.

Acknowledgements. Part of this work was supported by US Navy/Office of Naval Research Contract #N00014-11-M-0275 and NSF IIS Award #1149827.

The authors would also like to thank Sean Lyttle, Nicholas Pagano, Shi Bai and the fellows in MDL at Stevens Institute of Technology for the inspiring discussion and interdisciplinary collaboration on this work.

References

1. Donald, B.R., Levey, C.G., McGray, C.D., Paprotny, I., Rus, D.: An Untethered, Electrostatic, Globally Controllable MEMS Micro-Robot. Journal of Microelectromechanical Systems 15, 1–15 (2006)

2. Pac, M.R., Popa, D.O.: 3-DOF Untethered Microrobot Powered by a Single Laser Beam Based on Differential Thermal Dynamics. In: IEEE International Conference on Robotics and Automation, Shanghai (2011)
3. Sul, O.J., Falvo, M.R., Taylor, R.M., Washburn, S., Superfine, R.: Thermally actuated untethered impact-driven locomotive microdevices. Applied Physics Letters 89 (2006)
4. Yesin, K.B., Vollmers, K., Nelson, B.J.: Modeling and Control of Untethered Biomicrorobots in a Fluidic Environment Using Electromagnetic Fields. The International Journal of Robotics Research 25, 527–536 (2006)
5. Frutiger, D.R., Vollmers, K., Kratochvil, B.E., Nelson, B.J.: Small, Fast, and Under Control: Wireless Resonant Magnetic Micro-agents. The International Journal of Robotics Research 29, 613–636 (2009)
6. Pawashe, C., Floyd, S., Sitti, M.: Modeling and Experimental Characterization of an Untethered Magnetic Micro-Robot. The International Journal of Robotics Research 28, 1077–1094 (2009)
7. Hou, M.T., Shen, H.M., Jiang, G.L., Lu, C.N., Hsu, I., Yeh, J.A.: A rolling locomotion method for untethered magnetic microrobots. Applied Physics Letters 96 (2010)
8. http://www.femtotools.com
9. Jing, W., Pagano, N., Cappelleri, D.J.: A Micro-Scale Magnetic Tumbling Microrobot. In: Proceedings of the ASME International Design Engineering Technical Conferences & Computers and Information in Engineering, Chicago (2012)
10. Jing, W., Pagano, N., Cappelleri, D.J.: A novel micro-scale magnetic tumbling microrobot. Journal of Micro-Bio Robot 8, 1–12 (2013)
11. Jing, W., Pagano, N., Cappelleri, D.J.: A Tumbling Magnetic Microrobot with Flexible Operating Modes. In: IEEE International Conference on Robotics and Automation, Karlsruhe (2013)
12. Jing, W., Chen, X., Lyttle, S., Fu, Z., Shi, Y., Cappelleri, D.: Design of a magnetostrictive thin film microrobot. In: Proceedings of the ASME International Mechanical Engineering Congress & Exposition, Vancouver (2010)
13. Jing, W., Chen, X., Lyttle, S., Fu, Z., Shi, Y., Cappelleri, D.: A Magnetic Thin Film Microrobot with Two Operating Modes. In: IEEE International Conference on Robotics and Automation, Shanghai (2011)
14. Cappelleri, D.J., Piazza, G., Kummar, V.: A two dimensional vision-based force sensor for microrobotic applications. Sensors and Actuators A 171, 340–351 (2011)

Author Index